Robert Geroch

Geometrical Quantum Mechanics

1974 Lecture Notes

MINKOWSKI
Institute Press

Robert Geroch
Enrico Fermi Institute
University of Chicago

Cover: Lecture notes are often written in similar environments

ISBN: 978-1-927763-04-9 (softcover)
ISBN: 978-1-927763-03-2 (ebook)

Minkowski Institute Press
Montreal, Quebec, Canada
http://minkowskiinstitute.org/mip/

For information on all Minkowski Institute Press publications visit our website at http://minkowskiinstitute.org/mip/books/

Preface

This publication of Robert Geroch's course notes on geometrical quantum mechanics is the third volume in the new *Lecture Notes Series* of the *Minkowski Institute Press*. The idea of this series is to extend the life in space and time of valuable course notes in order that they continue to serve their noble purpose by bringing enlightenment to the present and future generations.

Geroch's lecture notes on geometrical quantum mechanics are divided into three parts – *Differential Geometry*, *Mechanics*, and *Quantum Mechanics*. The necessary geometrical ideas are presented in the first part of the book and are applied to mechanics and quantum mechanics in the second and third part. What also makes this book a valuable contribution to the existing textbooks on quantum physics is Geroch's unique approach to teaching theoretical and mathematical physics – the physical concepts and the mathematics, which describes them, are masterfully intertwined in such a way that both reinforce each other to facilitate the understanding of even the most abstract and subtle issues.

Robert Geroch would like to thank Rob Salgado for producing the initial LATEXversion of the typed course notes from mimeographed originals.

Montreal, March 2013 *Vesselin Petkov*
Minkowski Institute Press

Contents

Part I

Differential Geometry

1. Manifolds

The arena in which all the action takes place in differential geometry is an object called a *manifold*. In this section, we define a manifold, and give a few examples.

Roughly speaking, a n-dimensional manifold is a space having the "local smoothness structure" of Euclidean n-space, $\mathbb{R}^n$. (Euclidean n-space is the set consisting of n-tuples, $(x_1, \ldots, x_n)$, of real numbers). The idea, then, is to isolate, from the very rich structure of $\mathbb{R}^n$ (*e.g.*, its metric structure, vector-space structure, topological structure, etc.), that one bit of structure we call "smoothness".

Let M be a set. An n-**chart** on M consists of a subset U of M and a mapping $\psi : U \to \mathbb{R}^n$ having the following two properties:

1. The mapping ψ is one-to-one.

   (That is, distinct points of U are taken, by ψ, to distinct points of $\mathbb{R}^n$.)

2. The image of U by ψ, *i.e.*, the subset $O = \psi[U]$ of $\mathbb{R}^n$, is open in $\mathbb{R}^n$.

   (Recall that a subset O of $\mathbb{R}^n$ is said to be **open** if, for any point x of O, there is a number $\epsilon > 0$ such that the ball with center x and radius ϵ lies entirely within O.)

Let (U, ψ) be a chart. Then, for each point p of U, $\psi(p)$ is an n-tuple of real numbers. These numbers are called the **coordinates** of p (with respect to the given chart). The range of coordinates (*i.e.*, the possible values of $\psi(p)$ for p in U) is the open subset O of $\mathbb{R}^n$. By (1), distinct points of U have distinct coordinates. Thus, ψ represents n real-valued functions on U; a chart defines a labeling of certain points of M by real numbers.

These charts are the mechanism by which we intend to induce a "local smoothness structure" on the set M. They are well-suited to the job. Since a chart defines a correspondence between a certain set of points of M and a certain set of points of $\mathbb{R}^n$, structure on $\mathbb{R}^n$ can be carried back to U. To obtain a manifold, we must place on M a sufficient number of charts, and require that, when two charts overlap, the corresponding smoothness structures agree.

Let (U, ψ) and (U', ψ') be two n-charts on the set M. If U and U' intersect in M, there is induced on their intersection, $V = U \cap U'$, two "smoothness structures". We wish to compare them. To this end, we introduce the mapping $\psi' \circ \psi^{-1}$ from $\psi[V]$ to $\psi'[V]$ and its inverse $\psi \circ \psi'^{-1}$ from $\psi'[V]$ to $\psi[V]$. But

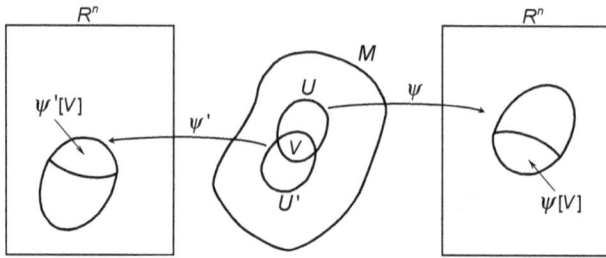

$\psi[V]$ and $\psi'[V]$ are subsets of $\mathbb{R}^n$, hence $\psi' \circ \psi^{-1}$ represents n functions-of-n-variables: $x'^1(x^1, \ldots, x^n)$, $\ldots, x'^n(x^1, \ldots, x^n)$. Similarly, $\psi \circ \psi'^{-1}$ represents the inverse functions: $x^1(x'^1, \ldots, x'^n)$, $\ldots$, $x^n(x'^1, \ldots, x'^n)$. These functions represent the interaction of (U, ψ) and (U', ψ') as regards smoothness structure on $V = U \cap U'$. We are thus led to the following definition: the n-charts (U, ψ) and (U', ψ') on M are said to be **compatible** if

1. $\psi[V]$ and $\psi'[V]$ are open subsets of $\mathbb{R}^n$

2. the mappings $\psi' \circ \psi^{-1}$ and $\psi \circ \psi'^{-1}$ are C^∞ (*i.e.*, all partial derivatives of these functions, of all orders, exist and are continuous).

Note that, for compatibility of charts, we require only that they agree in one structure of interest: *smoothness*. It is in this way that a single structure is isolated.

An n-**dimensional manifold** consists of a nonempty set M, along with a collection of n-charts on M such that

1. Any two charts in the collection are compatible.

2. Any chart on M which is compatible with all charts in the collection is also in the collection.

3. The charts in the collection cover M, *i.e.*, every point of M is in at least one of the charts.

4. The charts separate points of M, *i.e.*, if p and q are distinct points of M, then there are charts (U, ϕ) and (U', ϕ') in our collection such that p is in U, q is in U'; U and U' do not intersect.

These conditions—or at least the first three—are exactly what one might expect intuitively. The first condition states that *"whenever two charts induce competing smoothness structures on the same region of M, the structures agree"*. The second condition ensures that *we have "enough" charts*. The third condition ensures that *"smoothness structure is induced over all of M"*. The fourth condition eliminates pathological objects (called *non-Hausdorff* manifolds) which are of little interest.

One quickly gets an intuitive feeling for manifolds, so it becomes unnecessary constantly to return to these awkward definitions.

Example 1. Let M be the set $\mathbb{R}^n$, *i.e.*, the set of all n-tuples of real numbers. This M comes equipped already with a chart: set $U = M$, and ϕ the identity mapping (from M to $\mathbb{R}^n$). Now consider all n-charts on M which are compatible with this one. This set M, together with these charts, satisfies our four conditions for a manifold. (Condition 1 follows from the fact that smooth functions of smooth functions are smooth. The second and third conditions are obvious, and the fourth nearly so.) This n-dimensional manifold is called the manifold $\mathbb{R}^n$.

Example 2. Let M be the set of $(n + 1)$-tuples of real numbers, $(y^1, \ldots, y^{n+1})$, satisfying $(y^1)^2 + \cdots + (y^n)^2 = 1$. Denote by U the subset of M consisting of $(y^1, \ldots, y^{n+1})$ with $y^1 > 0$. Let ϕ denote the mapping from U to $\mathbb{R}^n$ defined by $x^1 = y^2$, $\ldots$, $x^n = y^{n+1}$. This is a chart on M. Similarly, introduce charts given by $y^1 < 0$, $y^2 > 0$, $y^2 < 0$, $\ldots$, $y^{n+1} < 0$. In this way, we obtain $2n + 2$ charts on M which, as can be easily checked, are compatible with each other. Now consider all charts on M which are compatible with these $2n + 2$ charts. As in Example 1, one can verify that this set M, with these charts, defines a manifold. It is called the manifold S^n, (the n-sphere).

Example 3. Let M be the set of all orthogonal 3×3 matrices (*i.e.*, matrices whose inverse is equal to their transpose). Let U consist of the subset M consisting of matrices are within $\frac{1}{10}$ of the corresponding entries in the unit-matrix. That is,

U consists of orthogonal matrices $\begin{pmatrix} a & b & c \\ d & e & f \\ g & h & i \end{pmatrix}$ with $|a - 1|$, $|e - 1|$, $|i - 1|$, $|b|$,

$|c|$, $|d|$, $|f|$, $|g|$, $|h|$ all less than $\frac{1}{10}$. For such a matrix K, set $\phi(K) = (b, c, f)$. This is a 3-chart on M. Fix any orthogonal matrix P. We define another chart, with U' consisting of matrices Q with QP in U, and with $\psi'(Q) = \psi(QP)$. Thus, for each P we obtain another chart. These charts are all compatible with each other. The collection of all charts on M compatible with these makes our set M into a 3-dimensional manifold. This is the manifold of the Lie-group $O(3)$. (Similarly, other matrix groups are manifolds).

2. Tensor Algebras

The natural things to exist on a manifold—the things which will describe the physics on a manifold—are objects called *tensor fields*. It is convenient to begin by ignoring manifolds completely. We shall define an abstract *tensor algebra*—a structure which makes no reference to a manifold. We shall then see, in the following section, that these tensor algebras arise naturally on manifolds.

A tensor algebra consists, first of all, of a certain collection of sets. We shall label these sets by a script "T" to which there is attached subscripts and superscripts from the Latin alphabet. In fact, we introduce a set for every arrangement of such subscripts and superscripts on $\mathscr{T}$, provided only that the arrangement is such that no index letter appears more than once. Thus, for example, our collection includes sets denoted by $\mathscr{T}^m{}_{rs}{}^b{}_a$, $\mathscr{T}_c$, $\mathscr{T}$, etc. The collection does not, however, include a set denoted by $\mathscr{T}_{dc}{}^{ac}{}_r$.

The elements of these sets will be called **tensors**. To indicate the particular set to which a tensor belongs, we shall write each tensor as a base letter with appropriate subscripts and superscripts. Thus, for example, $\xi_r{}^{pa}{}_{sd}$ represents a tensor, namely, an element of $\mathscr{T}_r{}^{pa}{}_{sd}$. Tensors with a single index (*e.g.*, η_c, an element of $\mathscr{T}_c$) are called **vectors**; tensors with no indices (e.g., α, an element of $\mathscr{T}$) **scalars**.

The next step in the definition of a tensor algebra is the introduction of four operations on these tensors.

1. Addition. With any two tensors which are elements of the same set (*i.e.*, which have precisely the same index structure), there is associated a third tensor, again an element of that same set. This operation will be written with a "+". Thus, $\lambda^c{}_{ad} + \omega^c{}_{ad}$ is an element of $\mathscr{T}^c{}_{ad}$.

2. Outer Product. With any two tensors having no index letter in common, there is associated a third tensor, an element of the set denoted by "$\mathscr{T}$" with first the indices of the first tensor, and then the indices of the second tensor attached. This operation will be indicated by writing the two tensors next to each other. Thus, for example, $\mu_{we}{}^r{}_t \nu^{yu}{}_c$ is an element of $\mathscr{T}_{we}{}^r{}_t{}^{yu}{}_c$.

3. Contraction. Given any tensor, along with a choice of a particular subscript and a particular superscript of that tensor, there is associated another tensor, an element of the set denoted by "$\mathscr{T}$" with all the indices of the original tensor, except the two chosen ones, attached. This operation is

indicated by writing the original tensor with the chosen subscript changed so as to be the same letter as the chosen subscript. Thus, for example, choosing the superscript "c" and the subscript "m" for $\xi_r{}^d{}_{mas}{}^{bc}$, we obtain by contraction the element of $\mathscr{T}_r{}^d{}_{as}{}^b$ written $\xi_r{}^d{}_{mas}{}^{bm}$. (Note that the final tensor is not an element of $\mathscr{T}_r{}^d{}_{mas}{}^{bm}$. There is no such set.)

4. Index Substitution. Given any tensor, together with a choice of an index of that tensor and a choice of a Latin letter which does not appear as an index of that tensor, there is associated another tensor, an element of the set denoted by "$\mathscr{T}$" with all the indices of the original tensor, except that the chosen index letter is replaced by the chosen letter. This operation is indicated by simply changing the appropriate index letter on the tensor. Thus, for example, if we choose the subscript "c" of $\eta^a{}_{cbn}$ and the Latin letter "r", the result of index substitution is $\eta^a{}_{rbn}$, an element of $\mathscr{T}^a{}_{rbn}$.

These four are the only (algebraic) operations available on tensors. What remains is to write the list of properties these operations satisfy.

1. *Conditions on addition.* Addition is associative and commutative. Within each set there is an additive identity (written "0", with indices suppressed), and each tensor has an additive inverse (written with a "−"). In short, each of our sets is an *abelian* group under addition.

$$(\alpha^a{}_b + \beta^a{}_b) + \gamma^a{}_b = \alpha^a{}_b + (\beta^a{}_b + \gamma^a{}_b)$$

$$\alpha^a{}_b + \beta^a{}_b = \beta^a{}_b + \alpha^a{}_b$$

$$\alpha^a{}_b + 0 = \alpha^a{}_b$$

$$\alpha^a{}_b + (-\alpha^a{}_b) = \alpha^a{}_b - \alpha^a{}_b = 0$$

2. *Conditions on outer product.* Outer product is associative and distributive over addition.

$$\mu^{ab}(\nu_c{}^m \tau_r) = (\mu^{ab}\nu_c{}^m)\tau_r$$

$$\mu^{ab}(\alpha_n{}^r + \beta_n{}^r) = \mu^{ab}\alpha_n{}^r + \mu^{ab}\beta_n{}^r$$

$$(\sigma_n{}^r + \tau_n{}^r)\gamma^{ab} = \sigma_n{}^r\gamma^{ab} + \tau_n{}^r\gamma^{ab}$$

3. *Conditions on contraction.* The operation of contraction commutes with all the operations. Thus, if $\alpha^a{}_b + \beta^a{}_b = \gamma^a{}_b$ then $\alpha^b{}_b + \beta^b{}_b = \gamma^b{}_b$. (On the right is the sum of two elements of $\mathscr{T}$; on the left, contraction applied to an element of $\mathscr{T}^a{}_b$.) If $\mu_{ab}{}^{crs} = \sigma_{ab}{}^c\tau^{rs}$, then $\mu_{ab}{}^{ars} = \sigma_{ab}{}^a\tau^{rs}$. If $\alpha^a{}_b = \kappa^a{}_b{}^d{}_d$ and $\beta^c{}_d = \kappa^b{}_b{}^c{}_d$, then $\alpha^b{}_b = \beta^d{}_d$. (This double contraction would be written $\kappa^b{}_b{}^d{}_d$.) The result of substituting "r" for "b" in $\lambda^c{}_{bc}$ is the same as the result of contracting $\lambda^a{}_{rc}$ over "a" and "c".

4. *Condition on index substitution.* The operation of index substitution commutes with all the operations. The result of substituting one index for another, followed by substituting the other for the one, is the original tensor.

These conditions are easy to remember because they are suggested by the notation.

5. The set of scalars, $\mathscr{T}$, is the set of real numbers, and addition and outer product for scalars is addition and multiplication of real numbers.

    It follows immediately that *each of our sets is actually a vector space (over the real numbers)*. Scalar multiplication is outer product by elements of $\mathscr{T}$. Furthermore, condition 4 implies that, *e.g.*, $\mathscr{T}^r{}_{ac}$ and $\mathscr{T}^s{}_{bn}$ are isomorphic vector spaces (the isomorphism obtained by index substitution). Furthermore, conditions 2 and 3 imply that the operation of contraction and outer product are linear mappings on the appropriate vector spaces.

6. Every tensor can be written as a (never unique) sum of outer products of vectors.

    On the outer product $\mu^a{}_{cdr}{}^s\alpha_u{}^v$, contract "s" with "u", and "v" with "r", to obtain an element of $\mathscr{T}^a{}_{cd}$ which may be written $\mu^a{}_{cdr}{}^s\alpha_s{}^r$. Thus, every element of $\mathscr{T}^a{}_{cdr}{}^s$ defines a linear mapping from the vector space $\mathscr{T}_s{}^r$ to the vector space $\mathscr{T}^a{}_{cd}$.

7. $\mathscr{T}^a{}_{cdr}{}^s$ consists precisely of the set of linear mappings from $\mathscr{T}_s{}^r$ to $\mathscr{T}^a{}_{cd}$, and similarly for other index combinations. Thus, for example, $\mathscr{T}_a$ is the dual of $\mathscr{T}^a$ (*i.e.*, the set of linear mappings from $\mathscr{T}$ to the reals). Further, $\mathscr{T}^a{}_b$ is the set of linear mappings from $\mathscr{T}^b$ to $\mathscr{T}^a$ (which, since $\mathscr{T}^b$ is isomorphic to $\mathscr{T}^a$, is the same as the set of linear mappings from $\mathscr{T}^b$ to $\mathscr{T}^b$). From this point of view, $\alpha^a{}_b\beta^b{}_c$ is the composition of a linear-mapping-from-$\mathscr{T}^c$-to-$\mathscr{T}^b$ with a linear-mapping-from-$\mathscr{T}^b$-to-$\mathscr{T}^a$ to obtain a linear-mapping-from-$\mathscr{T}^c$-to-$\mathscr{T}^a$. Further, $\alpha^a{}_a$ is the trace operation on linear mappings.

A **tensor algebra** is a collection of sets, on which four operations are defined, subject to seven conditions, as described above. We remark that the conditions are redundant. We have proceeded in this way in order to put the basic algebraic facts about tensors together on an equal footing.

The following central result summarizes much of linear algebra:

Theorem. Given any finite dimensional vector space V, there exists a tensor algebra with its $\mathscr{T}^a$ isomorphic to V. Furthermore, an isomorphism between the "$\mathscr{T}^a$'s" of two tensor algebras extends uniquely to an isomorphism of the tensor algebras.

In intuitive terms, a tensor algebra produces a framework for describing everything which can be obtained, starting from a single vector space, using linear mappings and tensor products. It equates naturally isomorphic things. It describes neatly the algebraic manipulations available on these objects.

Why didn't we require that outer product be commutative? Because it isn't defined. For example, $\alpha^{ab}\beta_c{}^d$ and $\beta_c{}^d\alpha^{ab}$ lie in different vector spaces, so equality is meaningless. We can, however, recover commutativity of outer product by introducing an additional convention. Let, for example, $\tau^a{}_{bc}$ be a tensor, and, by condition 6, write:

$$\tau^a{}_{bc} = \alpha^a \beta_b \gamma_c + \cdots + \mu^a \nu_b \kappa_c.$$

Then we can associate with this $\tau^a{}_{bc}$ an element, $\gamma_c \alpha^a \beta_b + \cdots + \kappa_c \mu^a \nu_b$, of $\mathscr{T}_c{}^a{}_b$. Thus, we obtain an isomorphism between $\mathscr{T}^a{}_{bc}$ and $\mathscr{T}_c{}^a{}_b$. Similarly, we have an isomorphism between any "$\mathscr{T}$" and any other obtained by changing the order of the indices, preserving their subscript or superscript status. We make use of these isomorphisms as follows. We permit ourselves to write equality of elements of different vector spaces, isomorphic as above, if the elements correspond under the isomorphism above. By this extended use of equality, for example, $\mu^a \nu_b = \nu_b \mu^a$. Similarly, we allow ourselves to add elements of isomorphic (as above) vector spaces by adding within one of the two spaces. Thus, $\gamma^a{}_{bc} = \delta_b{}^a{}_c + \sigma_{cb}{}^a$ means "carry $\sigma_{cb}{}^a$ to $\mathscr{T}_b{}^a{}_c$ via the isomorphism, and there add to $\delta_b{}^a{}_c$". The result, carried to $\mathscr{T}^a{}_{bc}$ via the isomorphism, is equal to $\gamma^a{}_{bc}$.

As a final example, let ξ_{ab} be a tensor. Use index substitution to successively obtain ξ_{ac}, ξ_{dc}, ξ_{da}, and ξ_{ba}. If it should happen that $\xi_{ab} = \xi_{ba}$, then our tensor is called **symmetric**. If it should happen that $\xi_{ab} = -\xi_{ba}$, then our tensor is called **antisymmetric**. In general, such a tensor will be neither symmetric nor antisymmetric. It can always be written uniquely as the sum of a symmetric and an antisymmetric tensor: $\xi_{ab} = \frac{1}{2}(\xi_{ab} + \xi_{ba}) + \frac{1}{2}(\xi_{ab} - \xi_{ba})$. These facts are familiar from matrix algebra.

3. Tensor fields

Our next task is to combine our discussion of manifolds with that of tensor analysis.

Let M be a manifold. A **scalar field** on M is simply a real-valued function on M. That is, a scalar field assigns a real number to each point of M. It turns out, however, that arbitrary scalar fields aren't very interesting: what is interesting are things we shall call *smooth* scalar fields. Let α be a scalar field on M. To define smoothness of α, we must make use of the charts on M. Thus, let (U, ψ) be a chart. Then the points of M in U are labeled, by this chart, by coordinates $(x^1, \ldots, x^n)$. Since α is just a function on M (and hence, also a function on U), we may represent α as one real function of the n variables $(x^1, \ldots, x^n)$. Formally, this function is $\alpha \circ \psi^{-1}$. We say that the scalar field is **smooth** if, for every chart (in the definition of the manifold M), $\alpha \circ \psi^{-1}$ (a function of n variables) is C^∞ (all partials of all orders exist and are continuous). We denote by $\mathscr{S}$ the collection of smooth scalar fields on M. It is obvious that sums and products of smooth scalar fields yield smooth scalar fields. It is also clear that there are many nonconstant smooth scalar fields on M.

We wish next to define the notion of a *vector on a manifold*. It is convenient, for motivation, to first recall some things about vectors in Euclidean space. A vector in Euclidean space can be represented by its components, $(\xi^1, \ldots, \xi^n)$. This representation is not, however, very convenient for a discussion of vectors on a manifold, because of the enormous freedom available on the choice of charts (with respect to which components could be taken) on a manifold. We would like, therefore, to think of some description of vectors in Euclidean space which refers less explicitly to components. Let $\alpha(x^1, \ldots, x^n)$ be a smooth function of n variables, *i.e.*, a smooth function on our Euclidean space. Then we can consider the directional-derivative of this function in the direction of our vector, evaluated at the origin:

$$\xi(\alpha) = \left(\xi^1 \frac{\partial \alpha}{\partial x^1} + \cdots + \xi^n \frac{\partial \alpha}{\partial x^n} \right)\bigg|_{x=0} \tag{1}$$

Thus, given a vector in Euclidean space, $\xi(\alpha)$ is a number for each function α.

It is immediate from the properties of partial-derivatives that $\xi(\alpha)$, regarded as a mapping from smooth functions on Euclidean space to the reals, satisfies the following conditions:

1. $\xi(\alpha + \beta) = \xi(\alpha) + \xi(\beta)$.

2. $\xi(\alpha\beta) = \alpha|_{x=0} \xi(\beta) + \beta|_{x=0} \xi(\alpha)$.

3. If $\alpha = \text{const}$, $\xi(\alpha) = 0$.

Thus, a vector in Euclidean space defines a mapping (namely, the directional derivative) from smooth functions on Euclidean space to the reals, satisfying the three conditions above. We now establish a sort of converse, to the general effect that these three conditions characterize vectors in Euclidean space. More precisely, we claim that:

A mapping from smooth functions on Euclidean space to the reals satisfying the three conditions above is of the form (1) for some $(\xi^1, \ldots, \xi^n)$.

Proof: Suppose we have such a mapping. Define n numbers by $\xi^1 = \xi(x^1)$, $\ldots$, $\xi^n = \xi(x^n)$. We shall show that, for this choice, (1) is valid for any smooth function α. Write such an α in the form

$$\alpha = \underline{\alpha} + \sum_i \alpha_i x^i + \sum_{ij} \alpha_{ij}(x) x^i x^j \tag{2}$$

where $\underline{\alpha}$ is a constant, the α_i $(i = 1, \ldots, n)$ are constants, and α_{ij} $(i, j = 1, \ldots, n)$ are smooth. Then, using conditions 1 and 2 above,

$$\xi(\alpha) = \xi(\underline{\alpha}) + \sum_i \left[\xi(\alpha_i) x^i + \alpha_i \xi(x^i) \right]$$
$$+ \sum_{ij} \left[\xi(\alpha_{ij}) x^i x^j + \alpha_{ij} \xi(x^j) x^i + \alpha_{ij} \xi(x^i) x^j \right].$$

Now use the condition 3 and evaluate at the origin:

$$\xi(\alpha) = \sum_i \alpha_i \xi(x^i) = \sum_i \alpha_i \xi^i$$

This is the left side of Eqn. (1). But, from (2), this is also clearly the right side of (1). Hence, (1) holds.

We have now completed our component-independent characterization of vectors in Euclidean space. This characterization is the motivation for the definition, which follows, of vectors on a manifold.

Fix a point p of our manifold M. A (**contravariant**) **vector** in M at p is a mapping $\xi : \mathscr{S} \to \mathbb{R}$ from the smooth scalar fields on M to the reals satisfying:

• $\xi(\alpha + \beta) = \xi(\alpha) + \xi(\beta)$.

• $\xi(\alpha\beta) = \alpha|_p \xi(\beta) + \beta|_p \xi(\alpha)$.

• If $\alpha = \text{const}$, then $\xi(\alpha) = 0$.

It is clear from the discussion above that the contravariant vectors at a point p of an n-dimensional manifold form an n-dimensional vector-space.

Intuitively, a contravariant vector at a point-of-a-manifold points in a "direction", just as do vectors in Euclidean space.

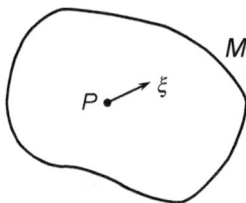

We now bring in the tensor algebras. At each point p of our manifold M, we have the n-dimensional vector-space of contravariant vectors at p. We now identify this vector space with the vector space $\mathscr{T}^a$ of a tensor algebra ("identify" means "make isomorphic to"). Then, as we have seen, we acquire a unique tensor algebra. These objects will be called *tensors* (in the manifold M) at the point p. Thus, a "scalar at p" is just a real number; a "vector ξ^a at p" is a contravariant vector, *i.e.*, a mapping from smooth scalar fields on M to reals satisfying our three conditions; a "vector η_a at p" is (among other things) a linear mapping from $\mathscr{T}^a$ at p to the reals. One repeats these remarks for each point of our manifold M. Thus, we have a notion of "a tensor at a point of M".

The next thing we must do is extend the notion of "a tensor at a point" to that of a tensor field. A **tensor field** on M assigns, to each point of M, a tensor at that point, where the tensors at the various points of M all have the same index-structure. For example, let $\alpha_{ac}{}^b{}_d{}^q$ be a tensor field on M. Then we have assigned, to each point of M, a tensor $\alpha_{ac}{}^b{}_d{}^q(p)$, at p. In particular, a scalar field on M, by this definition, assigns a scalar (*i.e.*, a real number) to each point of M. That is to say, a scalar field on M is just a real-valued function on M, which coincides with our original definition.

We observe that the operations defined within a tensor algebra extend from "tensors at a point" to tensor fields. Thus, let $\alpha_a{}^b$ and $\beta_a{}^b$ be tensor fields on M. Their sum, written as $\alpha_a{}^b + \beta_a{}^b$, is a tensor field on M which assigns, to the point p of M, the sum of the tensors assigned to p by $\alpha_a{}^b$ and $\beta_a{}^b$. That is, $(\alpha_a{}^b + \beta_a{}^b)(p) = \alpha_a{}^b(p) + \beta_a{}^b(p)$. Similarly, outer product, contraction, and index-substitution are well-defined operations on tensor fields. Furthermore, the various properties of tensors in a tensor algebra extend to properties of tensor fields. Thus, addition of tensor fields is associative and commutative. The zero tensor field (which assigns to each point the zero-tensor) is an additive identity. Outer products for tensor fields is associative and distributive. Contraction and index-substitution commute with everything in sight, and so on.

We saw a few pages ago that the notion of a scalar field is not very useful: what is useful is that of a *smooth* scalar field. This is perhaps not surprising, since the heart of a manifold is its smoothness structure, and one might expect that compatibility with that smoothness structure should be crucial. Since scalar fields were rejected in favor of smooth scalar fields, it is natural to ask whether tensor fields can somehow be rejected in favor of smooth tensor fields. This is indeed possible: we now do it.

Consider a contravariant vector field ξ^a. Let α be any smooth scalar field. Then, for each point p of our manifold, $\xi^a(p)$ is a contravariant vector at p.

Hence, it assigns, to the smooth scalar field α a number. Keeping ξ^a and α fixed, and repeating at each point p, we assign a real number to each point of M. In other words, we obtain a scalar field on M. (This scalar field is, of course, just the directional derivative of α in the direction of the vector field ξ^a.) We say that our contravariant vector field on M can be regarded as a mapping from smooth scalar fields on M to scalar fields on M. We say that our contravariant vector field ξ^a on M is **smooth** if it assigns, to each smooth scalar field on M a *smooth*-scalar field. Thus, we have extended the notion of smoothness from scalar fields to contravariant vector fields. A vector field η_a will be called *smooth* if, for each smooth ξ^a (smoothness for these just defined above), $\eta_a \xi^a$ is smooth. Finally, a tensor field, *e.g.*, $\kappa_{ac}{}^r{}_s{}^q$, will be called *smooth* if $\kappa_{ac}{}^r{}_s{}^q \xi^a \eta^c \lambda_r \omega^s \rho_q$ is a smooth scalar field for any smooth vector fields ξ^a, η^c, λ_r, ω^s, ρ_q. Note the way that the notion of smoothness permeates the tensor fields, starting from the scalar fields. This sort of thing is common in differential-geometry. We now notice that all the tensor operations, applied to smooth tensor fields, yield smooth tensor fields.

By the last sentence of the previous paragraph, if one starts with only smooth tensor fields, one never gets anything but smooth fields. It turns out in practice that it is just the smooth fields which are useful in physics. It is convenient, for this reason, not to have to always carry around the adjective "smooth". We shall henceforth omit it. When we say "tensor field", we mean "*smooth* tensor field" unless otherwise specified.

As we shall see shortly, manifolds and tensor fields are remarkably well-suited for the description of the arena in which physics takes place, and of the physics itself, respectively.

4. Derivative Operators

What is so great about smoothness? What was gained by our restricting consideration to smooth tensor fields? In the case of scalar fields, what we gained by requiring smoothness was the ability to take (directional) derivatives. This observation (together with the connotation of the word "smooth") suggests that it should be possible—and perhaps useful—to introduce the concept of the derivative of a (smooth) tensor field.

We can think of an n-dimensional manifold as similar to an n-dimensional surface. Thus, there are n directions in which one can move from a given point. That is, given a tensor field, there are n directions in which one can take the derivative of that tensor field. These remarks suggest that the operation "taking a derivative" should increase the number of indices of a tensor field by one. There is a more explicit way of seeing this. Let α be a scalar field. What should "the derivative of α" look like (as regards index-structure)? It would be right, somehow, if the derivative of α had one lowered index, $e.g.$, if it looked something like $\nabla_a \alpha$, for then $\xi^a \nabla_a \alpha$, for any contravariant vector field ξ^a, would naturally represent the directional derivative of α by ξ^a. The index on the "derivative operator" ∇_a provides a place to park the index of ξ^a, $i.e.$, it reflects the freedom in the direction in which the derivative can be taken. Thus, it seems reasonable, as a first try, to try define the "derivative operator" as an operator, with one down index, which acts on tensor fields. Having made this observation, it would be difficult to wind up with any definition for a derivative operator other than that which follows.

A **derivative operator**, ∇_a, is a mapping from (smooth) tensor fields to (smooth) tensor fields, which adds one lowered index, and which satisfies the following conditions:

1. The derivatives of the sum of two tensor fields is the sum of the derivatives. For example,
$$\nabla_a(\alpha^r{}_{cq} + \beta^r{}_{cq}) = \nabla_a(\alpha^r{}_{cq}) + \nabla_a(\beta^r{}_{cq}).$$

2. The derivative of the outer-product satisfies the Leibniz rule. For example,
$$\nabla_a(\mu^u{}_c{}^d \nu_{qv}) = \mu^u{}_c{}^d \nabla_a \nu_{qv} + \nu_{qv} \nabla_a \mu^u{}_c{}^d.$$

3. The derivative operation commutes with contraction and index substitution. For example, if $\beta = \alpha^a{}_a$ and $\gamma_c{}^a{}_b = \nabla_c \alpha^a{}_b$, then $\nabla_c \beta = \gamma_c{}^a{}_a$, and similarly for index substitution.

4. For any scalar field α and contravariant vector field ξ^a the scalar field $\xi^a \nabla_a \alpha$ is just the directional derivative.

5. Two derivative operations, applied to a scalar field, commute. That is, $\nabla_a \nabla_b \alpha = \nabla_b \nabla_a \alpha$. (Derivatives will <u>not</u> commute, in general, applied to tensor fields with more indices.)

Except possibly for the last, these are all natural-looking "derivative-like" conditions. It is easy to define things. What is usually more difficult is to show that these things exist, and to discover to what extent they are unique. We now ask these questions for derivative operators.

Every manifold one is ever likely to meet possesses a derivative operator. (Precisely, a manifold possesses at least one derivative operator if and only if it is paracompact.) In fact, it is a nontrivial job just to construct an example of a manifold which does not have any derivative operators. We shall henceforth deal only with manifolds which have at least one derivative operator. (In fact, this restriction will not be needed in any formal sense, but only to make discussions easier. We shall use the assumed existence of a derivative operator in three cases: to define Lie derivatives, exterior derivatives, and the derivative operator defined by a metric. Lie and exterior derivatives exist whether or not derivative operators exist. That the existence of a metric implies the existence of a derivative operator follows by a comparatively simple argument.)

A more interesting question is this: *how unique are derivative operators?* It turns out that derivative operators are never unique, but that the non-uniqueness can be expressed in a very simple way. We now proceed to explain this remark in more detail.

Suppose we had two derivative operators on the manifold M, ∇_a and ∇'_a. How are they related? We first consider the difference between their actions on a scalar field: $\nabla'_a \alpha - \nabla_a \alpha$. To draw conclusions about this expression, we contract it with an arbitrary contravariant vector field ξ^a, to obtain $\xi^a \nabla'_a \alpha - \xi^a \nabla_a \alpha$. This is the difference between two scalar fields. But, by condition 4, in the definition of a derivative operator, these two scalar fields are the same. So, the difference is zero. That is to say, $\xi^a (\nabla'_a \alpha - \nabla_a \alpha)$ vanishes for any ξ^a. Hence, $\nabla'_a \alpha - \nabla_a \alpha = 0$. That is, $\nabla'_a \alpha = \nabla_a \alpha$. In words, any two derivative operators have, by condition 4, the same action on scalar fields.

We next compare the actions of our two derivative operators on contravariant vector fields. Consider $\nabla'_a \xi^b - \nabla_a \xi^b$. We cannot conclude, as above, that this vanishes. However, it is true that this expression is linear in ξ^b. This follows from the following calculation

$$
\begin{aligned}
(\nabla'_a &- \nabla_a)(\alpha \xi^b + \eta^b) \\
&= \nabla'_a(\alpha \xi^b) - \nabla_a(\alpha \xi^b) + \nabla'_a(\eta^b) - \nabla_a(\eta^b) \\
&= \xi^b \nabla'_a(\alpha) + \alpha \nabla'_a(\xi^b) - \xi^b \nabla_a(\alpha) - \alpha \nabla_a(\xi^b) + (\nabla'_a - \nabla_a)\eta^b \\
&= \alpha(\nabla'_a - \nabla_a)\xi^b + (\nabla'_a - \nabla_a)\eta^b
\end{aligned}
$$

where, in the last step, we have used the fact that two derivative operators coincide on scalar fields. Hence, there is some tensor field $C^m{}_{ab}$ such that

$$\nabla'_a \xi^b - \nabla_a \xi^b = -C^b{}_{ac}\xi^c \qquad (3)$$

for all ξ^a. In short, derivative operators needn't be the same on contravariant vector fields, but their difference is expressible simply in terms of a certain tensor field.

Next, vector fields with one down index (covariant vector fields). Contract $\nabla'_a \eta_b - \nabla_a \eta_b$ with an arbitrary ξ^b, and use the properties of derivative operators:

$$
\begin{aligned}
\xi^b(\nabla'_a \eta_b - \nabla_a \eta_b) &= \nabla'_a(\eta_b \xi^b) - \eta_b \nabla'_a \xi^b - \nabla_a \xi^b \eta_b + \eta_b \nabla_a \xi^b \\
&= -\eta_b(\nabla'_a \xi^b - \nabla_a \xi^b) \\
&= -\eta_b(-C^b{}_{ac}\xi^c) \\
&= +\eta_m C^m{}_{ab}\xi^b.
\end{aligned}
$$

Since ξ^b is arbitrary, we have

$$\nabla'_a \xi^b - \nabla_a \xi^b = \eta_m C^m{}_{ab} \qquad (4)$$

Thus, the actions of our derivative operators on covariant vector fields also differ in a simple way, and, furthermore, the same tensor $C^m{}_{ab}$ as for contravariant vector fields comes into play.

We next derive, from the fifth property of derivative operators, a certain property of this $C^m{}_{ab}$. Let α be any scalar field. Then

$$
\begin{aligned}
\nabla'_a \nabla'_b \alpha - \nabla_a \nabla_b \alpha &= \nabla'_a \nabla_b \alpha - \nabla_a \nabla_b \alpha \\
&= (\nabla'_a - \nabla_a)\nabla_b \alpha = C^m{}_{ab}\nabla_m \alpha
\end{aligned}
$$

where, in the first step, we have used the fact that all derivative operators coincide on scalars. Now, the left side of this equation remains invariant if "a" and "b" are switched. So, the right side must. But this holds for all scalar fields α, and so

$$C^m{}_{ab} = C^m{}_{ba} \qquad (5)$$

What remains is to extend (3) and (4) to general tensor fields. This is done as follows. We wish to evaluate, say, $(\nabla'_a - \nabla_a)\alpha_{cd}{}^{rs}{}_u$. We contract it with vectors $\xi^c \eta^d \mu_r \nu_s \lambda^u$. We now differentiate by parts, just as we did in the calculation preceding Eqn. (4). The result is to shift the derivative operators to the vectors. Now use (3) and (4) to get rid of the derivative operators in favor of the $C^m{}_{ab}$. Finally, use the fact that the vectors are arbitrary—so they can be removed. The result (which, with a little thought, can be seen without actually doing the calculation) is

$$
\begin{aligned}
(\nabla'_a - \nabla_a)\alpha_{cd}{}^{rs}{}_u &= C^m{}_{ac}\alpha_{md}{}^{rs}{}_u + C^m{}_{ad}\alpha_{cm}{}^{rs}{}_u \\
&- C^r{}_{am}\alpha_{cd}{}^{ms}{}_u - C^s{}_{am}\alpha_{cd}{}^{rm}{}_u + C^m{}_{au}\alpha_{cd}{}^{rs}{}_m
\end{aligned} \qquad (6)
$$

Note the form of the right hand of Eqn. (6). There appears a term corresponding to each index of $\alpha_{cd}{}^{rs}{}_u$. Raised indices are treated just like ξ^b is in (3), while lowered indices are treated just like η_b is in (4). Thus, Eqns. (3) and (4), which are now special cases of (6), also make it easy to remember (6). Note also that the fact that all derivative operators coincide on scalar fields is a special case of Eqn. (6): no indices leads to no terms on the right.

We conclude that, given two derivative operators, ∇_a and ∇'_a, there is a tensor field $C^m{}_{ab}$ satisfying (5) such that the actions of these derivative operators are related by Eqn. (6). A converse is immediate. Suppose we have a derivative operator ∇_a, and suppose we choose any (smooth) tensor field $C^m{}_{ab}$ satisfying Eqn. (5). Then we can define a new operation, ∇'_a, on tensor fields by Eqn. (6). (for (6) can be interpreted now as giving the action of ∇'_a on any tensor field). It is immediate that this ∇'_a, so defined, satisfies the five conditions for a derivative operator. Thus, a derivative operator together with a $C^m{}_{ab}$ satisfying (5) defines another derivative operator.

In short, derivative operators are never unique, but we have good control over the non-uniqueness.

Part II

Mechanics

5. Configuration Space

The things in the physical world in which we shall be interested will be called *systems*. It is difficult, and probably futile, to try to formulate with any precision what a system is. Roughly speaking, a system is a collection of things (*e.g.,* objects, particles, fields) under study which, in some sense, do not interact with the rest of the universe. This splitting of the universe into separate systems one singles out for individual study is certainly arbitrary to some extent—that is to say, it is probably subject to what is fashionable in physics, and the mode of splitting probably changes as physics evolves. Nonetheless, we have to study something, and the things we study we call systems.

Examples of systems:

1. A free, point particle in Euclidean space.

2. A three-dimensional harmonic oscillator. That is, a point particle in Euclidean space, subject to a force directed towards the origin, and having strength proportional to the distance of the particle from the origin.

3. A point particle constrained to lie on a smooth two-dimensional surface S in Euclidean space.

4. A point particle in Euclidean space for which, for some reason, is unable to acquire any velocity component in the z-direction.

5. A ball free to roll (but not slide) over a horizontal plane in Euclidean space.

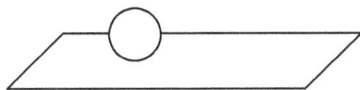

6. Electric and magnetic fields (say, without sources) in Euclidean space.

The systems listed above are not very exotic, and not very complicated. Neither are they described precisely.

The next step in the description of a system (after deciding what the system is) is the assignment to our system of a *configuration space*. The **configuration space**, C, is supposed to be a manifold, the points of which represent "configurations" of the system. There appears to be no definite rule, given an actual physical system (*e.g.*, a Swiss watch), for deciding what its space of configurations, C, should be. Apparently, this is essentially an art. For a simple mechanical system, the configurations are normally distinguished by what can be seen in an instantaneous photograph. Thus, in examples (1) and (2) above, one would normally choose for the set C the set of "locations of the particle", *i.e.*, the set of points in Euclidean 3-space. In example (3), one could, I suppose, also choose for C the set of points of Euclidean space. But that isn't a particularly good choice. One should instead, choose the "locations actually available to the particle", *i.e.*, the set of points on the surface S. It is not uncommon, in examples such as (3) above, to first make the "wrong" choice for configuration space, *i.e.*, to first choose Euclidean space. One then corrects oneself: he realizes that only certain of these configurations, namely, those on S, are actually available to the particle. This process—first choosing "too large" a configuration space, and then imposing restrictions—is called that of introducing *holonomic constraints*. It is my view that much is gained—and nothing lost—by being careful in the beginning. We shall be careful and, thus, we shall be able to avoid what are called holonomic constraints.

Configuration space is, as stated above, to be more than just a set of points—it is to be a manifold. That is to say, the set of configurations must be endowed with a smoothness structure, *i.e.*, with a collection of charts satisfying our four conditions for a manifold. Here, again, there appears to be no prescription, given a physical system and having decided what its set of configurations is to be, to determine the manifold structure. There is genuine content in this choice of smoothness structure: it, for example, tells us when two configuration are to be regarded as "close". In examples (1) and (2) above, one would naturally choose for the manifold structure on C the natural structure on Euclidean space. That is, in these examples, one would choose for C the manifold $\mathbb{R}^3$. In example (3), one would select charts appropriate to the embedding of S in Euclidean space. That is, in this example, C would be a 2-dimensional manifold (*e.g.*, a sphere, torus, plane, *etc.*, depending on what S is).

Note that the first two examples are essentially different physical systems, although they have the same configuration space.

Example (4) provides an illustration of these issues. Let us first choose for the set C the set of points in Euclidean space. What should we choose for the manifold structure? On the one hand, we could just choose the natural manifold structure on Euclidean space. Then C would be $\mathbb{R}^3$. There is, however, another reasonable option. A particle in this example remains in the same $z = $ constant plane all its life. Perhaps, then, we should regard each such plane as a separate "piece" of configuration space. Thus, perhaps we should endow C with the following manifold structure: C is the (disjoint) union of an infinite stack of 2-planes (each labeled by its z value). For this choice, C would be two-dimensional. For these two choices, we have the same set C, but different smoothness structures. There is even a third point of view. One could claim that the system hasn't been specified with enough care: that one should also be told the z-plane in which the particle beings. This additional information is to be regarded as part of the statement of what the system is. Then, of course, one would choose for C that one plane, *i.e.*, C would be $\mathbb{R}^2$ (since the only configurations available to the particle need be included in C). Thus, one could reasonably argue for any of the three configuration spaces for example (4): $\mathbb{R}^3$, the disjoint union of an uncountable collection of $\mathbb{R}^2$'s, or a simple $\mathbb{R}^2$. My taste leans mildly toward the second of these three choices.

To specify the "configuration of the system" in example (5), one must state where (on the plane) the sphere is, and the orientation of the sphere in space. "Where the plane is" would be represented by a point on the manifold $\mathbb{R}^2$, and the orientation (which could be described by the rotation necessary to obtain the desired orientation from some fixed orientation) by a point on the manifold O^3 (pp 5). Thus, a natural choice for the configuration space in this example would be the product of manifolds $\mathbb{R}^2 \times O^3$. We shall shortly define precisely the product of manifolds.)

Example (6) is a bit tricky. We are dealing, here, with vector fields $\vec{E}$ and $\vec{B}$ in Euclidean space, both of which have vanishing divergence. Here, the difficulty in giving a prescription for selecting configuration space becomes more clear. Perhaps the best one can do is try various possibilities for configuration space, and see which one works best. The one which has been found to work best is this: choose for C the set of all vector fields $\vec{B}$ in Euclidean space which have vanishing divergence. (Holonomic constraint types would say this: choose first all vector fields $\vec{B}$, and then restrict to the ones which have vanishing divergence.) This is the set C; what is the smoothness structure? It turns out that the only reasonable choice makes C what is called an *infinite-dimensional manifold*. We will not deal here with infinite-dimensional manifolds (because of technical complications). Hence, we shall not be able to deal in detail with example (6).

We summarize: reasonable people could disagree, given a physical system, on what configuration space is appropriate. For most systems, however, one choice is particularly natural, and so we shall allow to speak of the configuration space C of a system.

The dimension of the configuration space of a system is called the **number of degrees of freedom** of the system.

As times marches on, our system evolves, *i.e.*, it passes from one instant of time to the next, through various configurations. The mathematical description of this situation is via the notion of a curve. A **curve** on a manifold M is a mapping $\gamma : \mathbb{R} \to M$, where $\mathbb{R}$ is the reals. That is, for each real number t, $\gamma(t)$ is a point of M. So, as t sweeps through $\mathbb{R}$, $\gamma(t)$ describes our curve. A notion more useful is that of smooth curves, where smoothness for curves is defined in terms of smooth scalar fields. Let α be a smooth scalar field, so $\alpha : M \to R$. Then α. Then $\alpha \circ \gamma$ is a mapping from $\mathbb{R}$ to $\mathbb{R}$, intuitively, "the function α

evaluated, in terms of t, along the curve". We say that the curve γ is **smooth** if, for every smooth scalar field α, $\alpha \circ \gamma$ (one real function of one variable) is C^∞. Thus, the evolution of our system is described by a smooth curve γ, with parameter T representing time, on the configuration space C. Thus, whereas examples (1) and (2) have the same configuration space, the actual curves giving the evolution are different in the two examples.

We now need some more mathematics: the notion of the *tangent vector to a curve*. Let γ be a curve on M. Fix a number t_0, and set $p = \gamma(t_0)$. Now let α be a scalar field on M, and consider the number

$$\frac{d}{dt}\alpha(\gamma(t))\Big|_{t_0} .$$

Geometrically, this is the rate of change of the function α, with respect to t, along the curve, evaluated at t_0. It is immediate from the properties of the derivative (of functions of one variable) that this mapping from scalar fields on M to the reals satisfies the three conditions on page 12. Hence, this mapping defines a contravariant vector γ^a at p. This vector is called the **tangent vector** to the curve γ at the point $\gamma(t_0)$. Geometrically, the tangent vector is tangent to the curve, and is large when the curve "covers a lot of M for each increment of t".

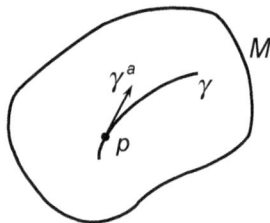

We now return to our system, its configuration space C, and some curve γ giving the evolution of the system. The tangent vector to this curve at $\gamma(t_0)$ is called the **velocity** of the system at time t_0, written v^a. This just generalizes the notion of the velocity, *e.g.*, of a particle moving in Euclidean space. Note that an evolving system, by our description, defines, at each moment of time, a configuration and a velocity.

As an illustration of the notion of velocity, we describe the set-up for *Lagrangian mechanics*. Let C be the configuration space for a system. The set of all pairs (q, v^a), where q is a point of C and v^a is a contravariant vector at q is called the **tangent bundle** of C. (This is technically not quite correct. The tangent bundle of a manifold is actually another manifold, having the dimension twice that of the original manifold. The set of pairs (q, v^a) above is actually just the underlying point-set of the tangent bundle.) The starting point for Lagrangian mechanics is the introduction of a certain function $L(q, v^a)$ on the tangent bundle of configuration space. This real-valued function is called the **Lagrangian** of the system. One sometimes writes the Lagrangian as $L(x, \partial x)$.

Are all points of the configuration space accessible to the system? In essentially all cases the answer is yes. In fact, that the answer be yes is one good criterion to use in selecting configuration space. Are all points of the tangent bundle accessible to the system? (More precisely, is it true that, given any point (q, v^a) of the tangent bundle, there exists a curve in configuration space, describing a possible evolution of the system, such that the curve passes through q and there has tangent vector v^a?) For many examples, the answer is yes, but it could be no.

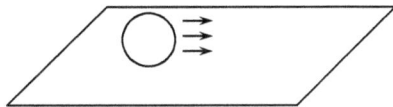

Consider example (5) on page 21. Consider any point of configuration space, and a contravariant vector which represents a translation of the sphere, without permitting it to roll. Here is a point of the tangent bundle which is not allowed to the system, because the prescribed contravariant vector would represent a velocity inconsistent with the requirement that the ball not slide on the plane. When it is the case that not every point of the tangent bundle is accessible to the system, one says that "the system exhibits a constraint". (One sometimes calls this a *nonholonomic constraint*. For us, the adjective is redundant, so we will not use it.) Thus, for example, for our three choices of the configuration space for example (4), the first has a constraint and the other two do not. There is certainly a sense in which the phrase "has a constraint" represents a property of the system itself more than a property of our description of the system. It appears to be difficult to make this remark precise. Does the point of the tangent bundle occupied by the system at one instant of time determine uniquely the future evolution of the system? (More precisely, do two curves, describing a possible evolution of the system, which pass through the same point of C for a given time t, and which there have the same tangent vector, necessarily coincide?) Again, examples suggest an answer yes, and this might be taken as a criterion for a good choice of configuration space. Again, this property is a rather complicated mixture of an aspect of Nature and an aspect we impose on Nature by our mode of description.

Suppose we have two systems, S_1 and S_2, and that we decide to regard them as a single system? (Note that nothing has happened physically: we just look at things differently.) If C_1 and C_2 are the configuration spaces of our systems, what should we choose for the configuration space of the combined system? The answer is $C = C_1 \times C_2$, a *product* of manifolds. We now define this. As a set, C consists of all pairs (p_1, p_2), where p_1 is a point of C_1 and p_2 is a point of C_2. We introduce charts. Let (U_1, ψ_1) be a chart on C_1, and (U_2, ψ_2) a chart on C_2. Consider the chart on C with $U = U_1 \times U_2$, and, for (p_1, p_2) in U, $\psi(p_1, p_2) = (\psi_1(p_1), \psi_2(p_2))$. (This last expression is an $(n_1 + n_2)$-tuple, where n_1 and n_2 are the dimensions of C_1 and C_2, respectively.) The collection of all charts of C compatible with these makes C a manifold. The dimension of C is, of course, $n_1 + n_2$. Thus, for example, the product of the manifold $\mathbb{R}^1$ and the manifold $\mathbb{R}^1$ is the manifold $\mathbb{R}^2$; The product of $\mathbb{R}^1$ and S^1 is the cylinder; The product of S^1 and S^1 is the torus. Thus, the combination of systems is represented, within our description, by the product of configuration space manifolds.

6. The Cotangent Bundle

It is convenient to interrupt briefly our study of mechanics at this point to introduce a little mathematics.

Let M be an n-dimensional manifold. We define, from M, a new $2n$-dimensional manifold Γ_M called the **cotangent bundle** of M. As a point-set, Γ_M consists of all pairs (q, p_a), where q is a point of M and p_a is a covariant vector in M at the point q. (It is not difficult to see, already at this point, why Γ_M will turn out to be $2n$-dimensional. It takes n dimensions to "locate" q in M, and, having chosen q, n more dimensions to locate p_a. (Recall that the vector space of covariant vectors at a point of an n-dimensional manifold is n dimensional.)) What remains is to introduce a collection of charts on this set Γ_M, and to verify that this collection of charts satisfies our four conditions for a manifold. Let (U, ψ) be any chart on M, so U is a subset of M and ψ, a mapping from U to $\mathbb{R}^n$. We associate, with this chart, a certain chart (U', ψ') (so U' will be a subset of Γ_M and ψ' a mapping from U' to $\mathbb{R}^{2n}$) on the set Γ_M. For U' we choose the collection of all pairs (q, p_a) with q in U. Then, for (q, p_a) in U', we set $\psi'(q, p_a) = (\psi(q), \kappa_1, \ldots, \kappa_n)$ where $\kappa_1, \ldots, \kappa_n$ are the n numbers such that

$$p_a = \kappa_1 \nabla_a x^1 + \cdots + \kappa_n \nabla_a x^n$$

where the right hand side is evaluated at q.

We now observe that these charts on Γ_M are all compatible with each other, and that they cover Γ_M. The collection of all charts on Γ_M compatible with these makes Γ_M a manifold, of dimension $2n$.

We repeat this construction in words. A chart on M labels points of M by n-tuples of real numbers. A point of Γ_M is a pair (q, p_a), where p_a is a covariant vector at the point q of M. Half the information needed to specify a point of Γ_M, namely q, is already labeled by n numbers, namely, the n numbers $\psi(q)$. For the other half, namely p_a, we observe that the gradients of the n coordinate functions define, at each point of U (in M), n covariant vectors which span the space of covariant vectors at that point. Hence, we can label p_a by the n numbers giving the expression for p_a, as a linear combination of the gradients of the coordinate functions. In this way, we obtain suitable charts on M.

There is a natural mapping, which we write π, from Γ_M to M. It is defined by $\pi(q, p_a) = q$. In words, π is the mapping which "forgets" what the covariant vector p_a is, but remembers the point q of M. A picture of this mapping, is that on the right. Each point q of M is located below a vertical line in Γ_M which represents all points of the form (q, p_a). Thus, in terms of this picture, π is the mapping which sends each point of Γ_M straight down to a point of M. It is immediate, for example, that π is always onto, and that it is one-to-one if and only if M is zero-dimensional.

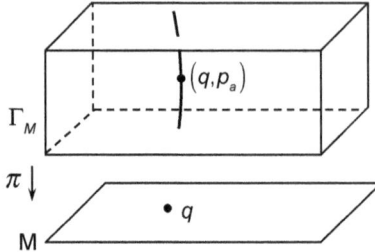

This π is our first example of a mapping from one manifold to another. It turns out that some such mappings are more interesting than others. The interesting ones are the ones we shall call *smooth*. Our next task is to define smoothness of a mapping from one manifold to another. Let $\tau : N \to M$ be a mapping from the manifold N to the manifold M. We say that this mapping is **smooth** if, for any smooth scalar field α on M, $\alpha \circ \tau$ (necessarily a real-valued function on N) is smooth. Note that $\alpha \circ \tau$ is the function on N whose value at a point x of N is the number assigned by α to the point $\tau(x)$ of M. One always defines smoothness of things on manifolds by asking that, when one does what one can with those things on smooth scalar fields, the result is smooth.

As an example of this notion of smoothness of mappings, we show that $\pi : \Gamma_M \to M$ is smooth. Let α be a smooth scalar field on M. Then $\beta = \alpha \circ \pi$ is the following function on Γ_M: $\beta(q, p_a) = \alpha(q)$. We must show that this function β is smooth. Let (U, ψ) be a chart on M, and let (U', ψ') be the corresponding chart on Γ_M. Then, in terms of the coordinates given by this chart, β is the following real function of $2n$ variables: $\beta(x^1, \ldots, x^n, \kappa_1, \ldots, \kappa_n) = \alpha(x^1, \ldots, x^n)$. Since α is smooth, the function (of n variables) on the right is smooth. Hence, the function (of $2n$ variables) representing β is smooth. Hence, β is a smooth scalar field on Γ_M. Hence, π is a smooth mapping.

The beautiful thing about the cotangent bundle of any manifold is that it possesses a natural covariant vector field. Since we are using Latin indices to label tensors on M, we shall, to avoid confusion, use Greek indices for tensors on Γ_M. This covariant vector field will be denoted A_α. Our next task is to define it.

Let (q, p_a) be a point of Γ_M (so, recall, q is a point of M and p_a is a covariant vector in M at q). Choose a scalar field α on M such that $\nabla_a \alpha$ (a covariant vector field on M) is, at this point q, just p_a. There are, of course, many such scalar fields on M; pick one. Then $\alpha \circ \pi$ is a scalar field on Γ_M. The gradient of this scalar field $\nabla_\beta(\alpha \circ \pi)$ is a covariant vector field on the manifold Γ_M. We evaluate this covariant vector field (on Γ_M) at the point (q, p_a) on Γ_M: this is A_β at the point (q, p_a) of Γ_M. Repeating for each point (q, p_a) of Γ_M, we obtain the desired vector field A_β on Γ_M. This A_β is the thing, in our formulation, which replaces "$p\ dx$" is classical mechanics.

To summarize, an n-dimensional manifold M gives rise, automatically, to a $2n$-dimensional manifold Γ_M, which is mapped smoothly to M, and on which there is a natural covariant vector field A_β.

7. Symplectic Manifolds

As in the previous section, let M be a manifold, and let Γ_M be the cotangent bundle of M. Then, as we have seen, there appears on Γ_M a certain vector field A_β. The curl of this vector field

$$\Omega_{\alpha\beta} = \nabla_\alpha A_\beta - \nabla_\beta A_\alpha \tag{7}$$

is called the **symplectic tensor field** of Γ_M. Which derivative operator should we use in (7)? Fortunately, this is not a decision we have to face, for the right hand side of (7) is independent of the derivative operator used. This is immediate from (6):

$$\nabla'_\alpha A_\beta - \nabla'_\beta A_\alpha = C^\mu{}_{\alpha\beta} A_\mu$$

where, in the last step we have used (5). The right hand side of (7) (except for the factor of two) is called the **exterior derivative** in differential-geometry.

The symplectic field will play a central role in our study of mechanics. The purpose of this section is to discuss its properties.

The most direct way of getting a quick feeling for the symplectic field is to express it in terms of chart. We first do this. Consider a chart on M, writing the coordinates $(x^1, \ldots, x^n)$. Then, as we saw in the previous section, we obtain a chart on Γ_M, with coordinates $(x^1, \ldots, x^n, \kappa_1, \ldots, \kappa_n)$. Now fix a point (q, p_a) of Γ_M, and let the coordinates of this point, with respect to our chart above, be $(\underline{x}^1, \ldots, \underline{x}^n, \underline{\kappa}_1, \ldots, \underline{\kappa}_n)$. We first express A_β, at the point (q, p_a), in terms of this chart on Γ_M. To do this, we just use the definition of A_β. Consider a scalar field α on M which, expressed in terms of our chart on M takes the form

$$\underline{\kappa}_1(x^1 - \underline{x}^1) + \ldots + \underline{\kappa}_n(x^n - \underline{x}^n). \tag{8}$$

It is immediate from the definition of the κ's (pp 27), that $\nabla_a \alpha$, evaluated at q, is just p_a. Furthermore, because of what π does, the scalar field $\alpha \circ \pi$, expressed in terms of our chart on Γ_M, is also given by (8). Hence, $A_\beta|_{(q,p)} = \nabla_\beta(\alpha \circ \pi)|_{(q,p)} = (\underline{\kappa}_1 \nabla_\beta x^1 + \ldots + \underline{\kappa}_n \nabla_\beta x^n)|_{(q,p)}$. Since the point (q, p_a) was arbitrary (except that it had to be in our chart on Γ_M), we have

$$A_\beta = \kappa_1 \nabla_\beta x^1 + \ldots + \kappa_n \nabla_\beta x^n. \tag{9}$$

It is eqn. (9) which suggests we interpret A_β as "$p\,dx$".

We now wish, similarly, to write the symplectic field in terms of these coordinates. This is easy: substitute (9) into (7). Doing this (and using the fact

31

that derivative operators commute when applied to a scalar field), we obtain immediately

$$
\begin{aligned}
\Omega_{\alpha\beta} = (\nabla_\alpha \kappa_1)(\nabla_\beta x^1) - (\nabla_\alpha x^1)(\nabla_\beta \kappa_1) + \ldots \\
+ (\nabla_\alpha \kappa_n)(\nabla_\beta x^n) - (\nabla_\alpha x^n)(\nabla_\beta \kappa_n)
\end{aligned}
\tag{10}
$$

This is the desired formula. In intuitive terms, (10) states that $\Omega_{\alpha\beta}$ has "q-p" components, and "p-q" components, of opposite sign, and that all "q-q" and "p-p" components of $\Omega_{\alpha\beta}$ vanish.

Finally, we introduce three fundamental properties of the symplectic field:

1. $\Omega_{\alpha\beta}$ is antisymmetric, *i.e.*, $\Omega_{\alpha\beta} = -\Omega_{\beta\alpha}$. This is immediate from the definition (7).

2. $\Omega_{\alpha\beta}$ is invertible, *i.e.*, there exists a (unique) antisymmetric tensor field $\Omega^{\alpha\beta}$ such that $\Omega_{\alpha\gamma}\Omega^{\beta\gamma} = \delta_\alpha{}^\beta$, where $\delta_\alpha{}^\beta$, the unit tensor, is defined by the property $\delta_\alpha{}^\beta \xi^\alpha = \xi^\beta$ for all ξ^α. (Invertibility for tensors is the tensor analog of invertibility for matrices.) This property follows, for example, from the explicit expression (10).

3. The curl of $\Omega_{\beta\gamma}$ vanishes, *i.e.*,

$$
\nabla_{[\alpha}\Omega_{\beta\gamma]} = 0
\tag{11}
$$

Here, and hereafter, square brackets, surrounding tensor indices, mean "write the expression with the enclosed indices in all possible orders, affixing a plus sign to those that are an even permutation of the original order, and a minus sign to those an odd permutation, and, finally, divide by the number of terms". Thus, in detail, the left hand side of (11) means $\frac{1}{6}(\nabla_\alpha\Omega_{\beta\gamma} + \nabla_\beta\Omega_{\gamma\alpha} + \nabla_\gamma\Omega_{\alpha\beta} - \nabla_\alpha\Omega_{\gamma\beta} - \nabla_\beta\Omega_{\alpha\gamma} - \nabla_\gamma\Omega_{\beta\alpha})$. The left hand side of (11) is in fact independent of the choice of derivative operator, an assertion proven exactly, as we proved the same assertion for (7). Finally, that (11) is indeed true can be seen, for example, by substituting (10). (Actually, (11) is a consequence of (7). $\nabla_{[\alpha}\nabla_\beta A_{\gamma]} = 0$ for any A_γ. To prove this, note that it is true that when $A_\gamma = \mu\nabla_\gamma V$, and that any A_γ can be expressed as a sum of such terms of this form.)

More generally, a **symplectic manifold** is a manifold (necessarily even dimensional) on which there is specified a tensor field $\Omega_{\alpha\beta}$ having the three properties above.

8. Phase Space

We return to mechanics. Let C be the configuration space of a system. The cotangent bundle of C, Γ_C, is called the **phase space** of the system. It was a consequence of our manner of introducing configuration space that our system, at any instant of time, occupies a particular point of configuration space. It will emerge shortly that our system may also be considered to occupy, at each instant, a point of phase space. That is to say, our system will "possess", at each instant, a pair (q, p_a), where p_a is a vector at point q of C. We shall continue to call q the configuration of the system. The covariant vector p_a at q will be called the **momentum** of the system.

Why does one proceed in this way? At first glance, one might think that the most natural thing to do would be to stay in configuration space. Points of configuration space have "concrete physical significance—they represent actual physical configurations of the actual physical system". Thus, the physics would somehow be closer to the surface in the mathematics if one could carry out the entire description directly in terms of the configuration space. However, a technical difficulty arises. As we pointed out in Sect. 5, knowing only the present configuration of a system does not suffice to determine what the system will do in the future. Roughly speaking, the configuration is half the necessary information. One might, therefore, think of going to the tangent bundle, of describing a system in terms of its configuration q and velocity v^a. Velocity is also, in a sense, "concrete", for one can look at a system (for a short span of time) and determine its velocity. Thus, by going to the tangent bundle, we could work with objects having direct physical significance, and, at the same time, deal with the $2n$ variables (q, v^a) necessary to determine the future behavior of the system.

In short, the tangent bundle seems ideal. Why, then, does one choose the cotangent bundle, the phase space, for the description of our system? The answer is not very satisfactory: because one has, on the cotangent bundle, the symplectic field. This field—this additional structure—seems to be necessary to write the equations which describe mechanics. We shall see that the symplectic field appears in almost every equation. There is nothing on the tangent bundle analogous to the symplectic field. The price one pays for the use of this field is a more tenuous connection between the mathematics and the actual physics. Suppose, for example, that we have an actual physical system, and that we have assigned to it some configuration space C. The system is set into action, and we are permitted to watch the system. We thus obtain a curve in configuration space. What is the momentum of this system at some instant of time? There is no definite way to tell (and, in fact, as we shall see later, there is some ambiguity in this assignment). In a sentence, the momentum of a system is a pure kinematical quantity. On the other hand, the cotangent bundle (phase space) has the same dimension as the tangent bundle. We replace a concrete thing, velocity, by a more abstract thing, momentum, in order to acquire the symplectic field for use in equations. What is unsatisfying is that it is not clear why Nature prefers additional structure over a direct physical interpretation.

Points of phase space will be called **states** of the system. Thus, to give the state of a system, one must specify its configuration

9. The Hamiltonian

The configuration space concerns *kinematics*—what can happen. The phase space is an introduction to the present section, in which we discuss *dynamics*—what actually does happen.

At each instant of time, the system is to possess a state, *i.e.*, a point of phase space, *i.e.*, a configuration and momentum. The evolution of the system is thus described by a curve, $\gamma : \mathbb{R} \to \Gamma_C$, in phase space. Thus, for example, $\pi \circ \gamma$ is a curve in configuration space, a curve which describes the sequence of configurations through which the system passes. Thus, dynamics is to be described by giving a bunch of curves in phase space. Since phase space is $2n$-dimensional, one would expect that precisely one of these curves will pass through each point of phase space. The required statement is this: there is a (smooth) scalar field H on phase space, called the **Hamiltonian**, such that the evolution is described by curves in phase space with tangent vector

$$H^\alpha = \Omega^{\beta\alpha} \nabla_\beta H \tag{12}$$

This vector field H^α on phase space is called the **Hamiltonian vector field**. (There is a theorem in differential-geometry to the effect that, given a contravariant vector field on a manifold, there passes, through each point of that manifold, precisely one curve everywhere tangent to that vector field.)

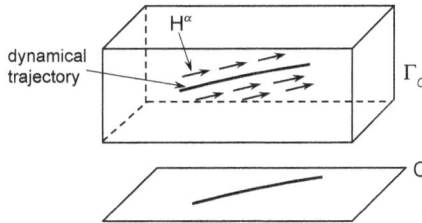

That the statement above has, essentially, the content of **Hamilton's equations** ($\dot{q} = \frac{\partial H}{\partial p}$, $\dot{p} = -\frac{\partial H}{\partial q}$) is clear from the explicit expression (10) for the symplectic tensor field.

The curves in phase space everywhere tangent to the Hamiltonian vector field in phase space will be called the **dynamical trajectories**. The fundamental statement of dynamics can now be formulated as follows: physical systems in Nature seem often to have the property that, when an appropriate configuration space is selected, there exists a Hamiltonian H, a scalar field on phase space,

35

such that the corresponding dynamical trajectories, when projected down to configuration space, give the actual sequence of configurations through which the system passes with time.

Example: Suppose that our physical system is a free particle (say, to keep things simple, in one dimension) of mass m. One could choose for the Hamiltonian $H = \frac{p^2}{2m}$, and obtain the equations of motion $\dot{x} = \frac{p}{m}$, $\dot{p} = 0$. However, one could as well choose for the Hamiltonian $H = (p + x^2)^2$. Then the equations of motion would be $\dot{x} = 2(p + x^2)$, $\dot{p} = -4x(p + x^2)$. All one actually sees about the physical system is $\ddot{x} = 0$. But this is the case for either Hamiltonian. In particular, if one is given x and $\dot{x}$ for the system at one instant of time, the momentum (and hence the point of phase space) one assigns to the system at that time depends on which Hamiltonian one has chosen. The momentum, in short, is "partially a kinematic and partially a dynamic quantity".

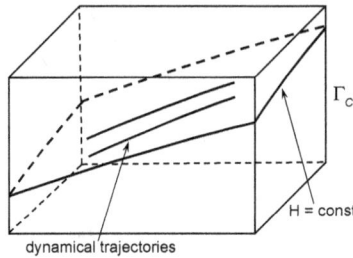

dynamical trajectories

Finally, we show that the scalar field H is constant along the dynamical trajectories. The rate of change of the Hamiltonian along a dynamical trajectory is the directional derivative of H in the direction of the tangent vector to the curve, i.e., the directional derivative of H along the Hamiltonian vector field. But, from (12), $H^{\alpha} \nabla_{\alpha} H = (\Omega^{\beta\alpha} \nabla_{\beta} H)(\nabla_{\alpha} H) = 0$, where the last step is a consequence of the antisymmetry of $\Omega^{\beta\alpha}$. We can represent this fact geometrically. Draw, in phase space, the $(2n - 1)$-dimensional surfaces $H = $ constant. These surfaces fill phase space. The remark above states that each dynamical trajectory remains forever within a single $H = $ constant surface.

10. Observables; Poisson Bracket

By an **observable** of a system, we mean a scalar field on its phase space. Roughly speaking, observables are the things instruments measure. We think of an observable as a box having a dial and a little probe which sticks into the system. The probe is sensitive to what the system is doing, and thus causes the needle to move to some point on the dial. It should be noted, however, that since, given the actual system sitting there, the momentum one assigns to it may depend on the choice (possibly arbitrary to some extent) of a Hamiltonian, one must think of a scalar field on phase space as being associated, not only with the actual instrument which probes the system, but also with our mode of description of the system. Nevertheless, one supposes that the configuration space—phase space—Hamiltonian assignments have been made, somehow, and calls a scalar field on phase space an observable.

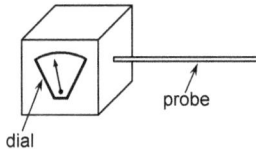

Two particular types of observables are of particular interest. Let α be a scalar field on configuration space. Then $\alpha \circ \pi$ is a scalar field on phase space (which ignores momentum, and looks only at configuration). Any observable of this form, $\alpha \circ \pi$, will be called a **configuration observable**. In elementary treatments of particle mechanics, "x", "y", and "z" are configuration observables. Now consider a contravariant vector field ξ^a on configuration space. We introduce the following scalar field on phase space: it assigns, to the point (q, p_a) of phase space, the number obtained by contracting ξ^a at q (a contravariant vector at q) with p_a (a covariant vector at q). We write this scalar field $\xi^a p_a$. An observable expressible in this form will be called a **momentum observable**. In elementary particle mechanics, "p_x", "p_y", and "p_z" are examples of momentum observables (*i.e.*, those defined by unit vectors pointing in the x, y, and z directions). Clearly, some observables (*i.e.*, configuration observables) are "more observable" than others (*e.g.*, momentum observables).

What operations are available on observables? One can certainly add them (addition of scalar fields) and multiply them (outer product of scalar fields).

However, there is another, equally important, operation called *Poisson bracket*. If A and B are observables, the **Poisson bracket** of A and B (a third observable) is defined by

$$\{A, B\} = \Omega^{\alpha\beta}(\nabla_\alpha A)(\nabla_\beta B) \tag{13}$$

Note how the symplectic field serves to "mix configuration and momentum components", so that this expression is the same as the familiar one.

There are five basic properties of Poisson brackets:

1. Additive: $\{A+B, C\} = \{A, C\} + \{B, C\}$. This is immediate from (13) and from the fact that the derivatives of (scalar) fields are additive.

2. Leibniz rule: $\{AB, C\} = A\{B, C\} + \{A, C\}B$. This is immediate from the Leibniz rule for derivatives of (scalar) fields.

3. Vanish for constants: If $A = $ constant, then $\{A, B\} = 0$. This is immediate from the fact that the derivative of a constant scalar field is zero.

4. Antisymmetry: $\{A, B\} = -\{B, A\}$. This is immediate from (13) and the fact that $\Omega^{\alpha\beta}$ is antisymmetric.

5. Jacobi identity: $\{A, \{B, C\}\} + \{B, \{C, A\}\} + \{C, \{A, B\}\} = 0$. This is not so immediate. We have

$$\begin{aligned}
\{A, \{B, C\}\} &= \Omega^{\alpha\beta}(\nabla_\alpha A)(\nabla_\beta[\Omega^{\gamma\delta}(\nabla_\gamma B)(\nabla_\delta C)\,]\,) \\
&= \Omega^{\alpha\beta}[\nabla_\beta\Omega^{\gamma\delta}](\nabla_\alpha A)(\nabla_\gamma B)(\nabla_\delta C) \\
&\quad + \Omega^{\alpha\beta}\Omega^{\gamma\delta}\nabla_\alpha A([\nabla_\beta\nabla_\gamma B]\nabla_\delta C + \nabla_\gamma B[\nabla_\beta\nabla_\delta C])(14)
\end{aligned}$$

where, in the first step, we have substituted (13), and, in the second, we have expanded using the Leibniz rule for a derivative operator. Now suppose we add the right side of (14) to itself three times, switching, each time, the order of A, B, and C as described by the Jacobi identity. Then the second term on the far right in (14) will cancel with itself. Hence, the Jacobi identity will be proven if we can show

$$\Omega^{\beta[\alpha}\nabla_\beta\Omega^{\gamma\delta]} = 0 \tag{15}$$

because this fact would suffice to kill off the first term on the far right in (14). Since $\Omega_{\alpha\beta}$ is invertible, (15) is equivalent to

$$\Omega_{\mu\alpha}\Omega_{\nu\gamma}\Omega_{\sigma\delta}\Omega^{\beta[\alpha}\nabla_\beta\Omega^{\gamma\delta]} = 0 \tag{16}$$

Now differentiate by parts (*i.e.*, use the Leibniz rule for derivative operators):

$$\begin{aligned}
&\Omega_{\mu\alpha}\Omega_{\nu\gamma}\Omega_{\sigma\delta}\Omega^{\beta\alpha}\nabla_\beta\Omega^{\gamma\delta} \\
&= \Omega_{\mu\alpha}\Omega_{\nu\gamma}\Omega^{\beta\alpha}\nabla_\beta(\Omega_{\sigma\delta}\Omega^{\gamma\delta}) - \Omega_{\mu\alpha}\Omega_{\nu\gamma}\Omega^{\beta\alpha}\Omega^{\gamma\delta}\nabla_\beta\Omega_{\sigma\delta} \\
&= +\nabla_\mu\Omega_{\sigma\nu}
\end{aligned}$$

Finally, using the fact that $\nabla_{[\alpha}\Omega_{\beta\gamma]} = 0$, we obtain (15), and hence the Jacobi identity.

The five properties of the Poisson bracket are easy to remember: the first three are properties of the derivative of scalar fields, the last two are properties of the symplectic field.

Finally, we derive the well-known equation for the rate of change with time of an observable. Let A be an observable. Then the time rate of change of A (along a dynamical trajectory) is (by definition of tangent vector) the directional derivative of A in the direction of the tangent vector to this dynamical trajectory. But this tangent vector is the Hamiltonian vector field. Hence,

$$\dot{A} = H^\alpha \nabla_\alpha A = \Omega^{\beta\alpha}(\nabla_\beta H)(\nabla_\alpha A) = \{H, A\} \tag{17}$$

Note, in particular, that, if we replace "A" by "H" in (17), we obtain, by anti-symmetry of the Poisson bracket, $\dot{H} = 0$. We have already seen this.

11. Canonical Transformations

We begin with a little mathematics. Let M and N be manifolds and let $\psi :$ $M \to N$ be a mapping. We say that this ψ is a **diffeomorphism** if

1. ψ is a smooth mapping,

2. ψ^{-1}, a mapping from N to M, exists (*i.e.*, ψ is one-to-one and onto) and is also a smooth mapping.

Intuitively, a diffeomorphism between M and N makes "M and N identical manifolds". Diffeomorphisms are to manifolds as isomorphisms are to vector spaces, or homeomorphisms are to topological spaces.

The remark above suggests that a diffeomorphism from M to N should be prepared to carry tensor fields from M to N and back. This in fact is the case. Let α be a scalar field on M. Then $\alpha \circ \psi^{-1}$ is a scalar field on N. It is sort of the "image" on N of α on M; we write this scalar field $\psi(\alpha)$. We now know how to carry scalar fields from M to N. Since tensor fields are built up in terms of scalar fields, it should now be possible to carry tensor fields from M to N. Indeed, consider the following equations:

$$\psi(\xi^a \nabla_a \alpha) = \psi(\xi^a) \nabla_a \psi(\alpha) \tag{18}$$

$$\psi(\xi^a \eta_a) = \psi(\xi^a) \psi(\eta_a) \tag{19}$$

$$\psi(T^a{}_c{}^{rs}{}_d \xi_a \eta^c \lambda_r \omega_s \kappa^d) = \psi(T^a{}_c{}^{rs}{}_d) \psi(\xi_a) \psi(\eta^c) \psi(\lambda_r) \psi(\omega_s) \psi(\kappa^d) \tag{20}$$

Eqns. (18), (19), and (20) are trying to say that tensor operations commute with the action of ψ. But we wish to use them as the definition of ψ. Eqn. (18) defines $\psi(\xi^a)$, *i.e.*, it defines a mapping from contravariant vector fields on M to contravariant vector fields on N. In words, the definition (18) is this: if ξ^a is a contravariant vector field on M, $\psi(\xi^a)$ is that contravariant vector on N which, applied (via directional derivative) to a scalar field $\psi(\alpha)$ on N, yields the same result as that of applying ξ^a to α (on M), and carrying the result over (via ψ) to N. In short, (18) requires that the operation "directional derivative" be invariant under the pushing of fields from M to N. Similarly, (19) defines $\psi(\eta_a)$, *i.e.*, (19) defines a mapping from covariant vector fields on M to covariant vector fields on N. This definition amounts to requiring that contraction be invariant under the "pushing action" of ψ. Finally, (20) defines a mapping from arbitrary tensor fields on M to fields on N.

In short, a diffeomorphism from one manifold to another provides a mechanism for carrying tensor fields from the first manifold to the second. This mechanism is completely and uniquely determined by the following properties: on scalar fields, it is the obvious thing, and it commutes with all tensor operations.

Now consider a system, with configuration space C, phase space Γ_C, symplectic structure $\Omega_{\alpha\beta}$, and Hamiltonian H. A diffeomorphism $\psi : \Gamma_C \to \Gamma_C$ from phase space to itself which leaves invariant the symplectic field,

$$\psi(\Omega_{\alpha\beta}) = \Omega_{\alpha\beta} \qquad (21)$$

is called a **canonical transformation**. That is to say, canonical transformations are mappings from phase space to itself which preserve all of the structure of interest there (smoothness, symplectic). Note that canonical transformations need not leave the Hamiltonian invariant. When this is the case, *i.e.*, when

$$\psi(H) = H \qquad (22)$$

the canonical transformation will be called a **symmetry**. For a moment, regard the phase space, and symplectic field as representing "kinematics", and these things, together with the Hamiltonian, as representing "dynamics". Then the following sentence is both a motivation for, and a summary of, the definitions above: canonical transformations are to kinematics as symmetries are to dynamics as isomorphisms are to vector spaces.

Because they are difficult to manipulate, one seldom uses canonical transformations in practice. Instead, one introduces what are called *infinitesimal canonical transformations*. In order to define infinitesimal canonical transformations, we must, again, introduce a few properties of manifolds.

Let M be a manifold. Consider the identity diffeomorphism on M, the one that takes each point of M to itself. Then, roughly speaking, an infinitesimal diffeomorphism on M should be the one which is near the identity, *i.e.*, one that takes each point of M to a nearby point. But a contravariant vector field on M defines a "direction of motion in M" at each point. These remarks suggest that a contravariant vector field on M is the precise geometrical object appropriate to replace the intuitive notion of an "infinitesimal diffeomorphism". A diffeomorphism from M to M carries tensor fields on M to tensor fields on M. One would expect, therefore, that an infinitesimal diffeomorphism on M carries tensor fields on M to nearby tensor fields on M. The appropriate thing to consider, therefore, is the "infinitesimal change in the tensor fields under infinitesimal diffeomorphism". The precise representation of an "infinitesimal diffeomorphism" on M is a contravariant vector field on M. This remark suggests that the precise representation of "the infinitesimal change in a tensor field under an infinitesimal diffeomorphism on M" should be something like "the directional derivative of a tensor field in the direction of ξ^a". The remarks above are merely motivation for the definitions which follow.

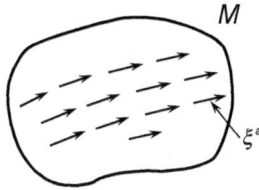

Fix a contravariant vector field ξ^a on the manifold M. The symbol $\mathcal{L}_\xi$ will be called the **Lie derivative** (in the direction of ξ^a). For α, a scalar field on M, set

$$\mathcal{L}_\xi \alpha = \xi^a \nabla_a \alpha. \tag{23}$$

Thus, the Lie derivative of a scalar field is the directional derivative. As before, we now wish to extend the notion of the Lie derivative from scalar fields to tensor fields. Consider the following equations:

$$\mathcal{L}_\xi(\eta^a \nabla_a \alpha) = (\mathcal{L}_\xi \eta^a)\nabla_a \alpha + \eta^a \nabla_a(\mathcal{L}_\xi \alpha) \tag{24}$$

$$\mathcal{L}_\xi(\eta^a \lambda_a) = (\mathcal{L}_\xi \eta^a)\lambda_a + \eta^a \mathcal{L}_\xi \lambda_a \tag{25}$$

$$
\begin{aligned}
\mathcal{L}_\xi(T^{ac}{}_d \eta_a \lambda_c \kappa^d) = {}& \mathcal{L}_\xi(T^{ac}{}_d)\eta_a \lambda_c \kappa^d + T^{ac}{}_d \mathcal{L}_\xi(\eta_a)\lambda_c \kappa^d \\
& + T^{ac}{}_d \eta_a \mathcal{L}_\xi(\lambda_c)\kappa^d + T^{ac}{}_d \eta_a \lambda_c \mathcal{L}_\xi(\kappa^d)
\end{aligned} \tag{26}
$$

These equations are trying to say that the Lie derivative satisfies the "Leibniz rule, as any good directional derivative type operator should do". But we use them as definitions. Eqn. (24) is the definition of the Lie derivative of a contravariant vector field (η^a) on M; Eqn. (25) the definition of the Lie derivative of a covariant vector field on M; Eqn. (26) the definition of the Lie derivative of an arbitrary tensor field on M. To summarize, the Lie derivative, "the directional derivative generalized to tensor fields", is completely and uniquely characterized by the following properties: on scalar fields, it is the obvious thing, and it satisfies Leibniz-type rules.

One actually uses in practice, not the definitions above for the Lie derivative, but more explicit expressions we now derive. We have

$$\xi^b \nabla_b(\eta^a \nabla_a \alpha) = \mathcal{L}_\xi \eta^a (\nabla_a \alpha) + \eta^a \nabla_a(\xi^b \nabla_b \alpha)$$

$$(\xi^b \nabla_b \eta^a)\nabla_a \alpha + \xi^b \eta^a \nabla_b \nabla_a \alpha = (\mathcal{L}_\xi \eta^a)\nabla_a \alpha + (\eta^a \nabla_a \xi^b)\nabla_b \alpha + \eta^a \xi^b \nabla_a \nabla_b \alpha$$

$$(\xi^b \nabla_b \eta^a)\nabla_a \alpha = (\mathcal{L}_\xi \eta^a)\nabla_a \alpha + (\eta^a \nabla_a \xi^b)\nabla_b \alpha$$

where the first equation results from using (23) in (21), the second equation results from expanding using the Leibniz rule (for a derivative operator), and the third equation results from the fact that derivatives commute on scalar fields. Thus, since α is arbitrary in the last equation above, we have

$$\mathcal{L}_\xi \eta^a = \xi^b \nabla_b \eta^a - \eta^b \nabla_b \xi^a \tag{27}$$

We now, similarly, use (23) and (27) in (25)

$$\xi^b \nabla_b(\eta^a \lambda_a) = (\xi^b \nabla_b \eta^a - \eta^b \nabla_b \xi^a)\lambda_a + \eta^a \pounds_\xi \lambda_a$$

$$(\xi^b \nabla_b \eta^a)\nabla_c \alpha + \xi^b \eta^a \nabla_b \nabla_a \alpha = (\pounds_\xi \eta^a)\nabla_a \alpha + (\eta^a \nabla_a \xi^b)\nabla_b \alpha + \eta^a \xi^b \nabla_a \nabla_b \alpha$$

$$(\xi^b \nabla_b \eta^a)\nabla_a \alpha = (\pounds_\xi \eta^a)\nabla_a \alpha + (\eta^a \nabla_a \xi^b)\nabla_b \alpha$$

Since η^a is arbitrary,

$$\pounds_\xi \lambda_a = \xi^b \nabla_b \lambda_a + \lambda_b \nabla_a \xi^b \tag{28}$$

Finally, using (27) and (28) in (26), we obtain, in precisely the same way above,

$$\begin{aligned}\pounds_\xi T^{ab}{}_{cd} &= \xi^m \nabla_m T^{ab}{}_{cd} - T^{mb}{}_{cd}\nabla_m \xi^a - T^{am}{}_{cd}\nabla_m \xi^b \\ &\quad + T^{ab}{}_{md}\nabla_c \xi^m + T^{ab}{}_{cm}\nabla_d \xi^m\end{aligned} \tag{29}$$

We could just as well have defined the Lie derivative operation by (29). (Note that (23), (27), and (28) are just special cases, and that (24), (25), and (26) follow immediately from (29).) Eqn. (29) is easily the most useful expression for the Lie derivative. Finally, we note that the right hand side of (29) is in fact independent of the choice of derivative operator. This is obvious from our definition, but can also be checked directly from (6).

We now again return to our system, with its configuration space, phase space, symplectic structure, and Hamiltonian. An **infinitesimal canonical transformation** on the system is a contravariant vector field ξ^a on phase space, Γ_C, such that

$$\pounds_\xi \Omega_{\alpha\beta} = 0 \tag{30}$$

An **infinitesimal symmetry** is an infinitesimal canonical transformation ξ^a which, in addition, satisfies

$$\pounds_\xi H = 0 \tag{31}$$

Everything in this section so far, except for (29) and these two definitions, can be regarded as motivation.

There is a crucial formula, which simplifies the discussion of infinitesimal canonical transformations, namely,

$$3\xi^\gamma \nabla_{[\alpha}\Omega_{\beta\gamma]} - \pounds_\xi \Omega_{\alpha\beta} - 2\nabla_{[\alpha}(\Omega_{\beta]\gamma}\xi^\gamma) = 0 \tag{32}$$

To prove (32), we simply expand it

$$\begin{aligned}\xi^\gamma \nabla_\alpha \Omega_{\beta\gamma} + \xi^\gamma \nabla_\beta \Omega_{\gamma\alpha} + \xi^\gamma \nabla_\gamma \Omega_{\alpha\beta} \\ -\xi^\gamma \nabla_\gamma \Omega_{\alpha\beta} - \Omega_{\gamma\beta}\nabla_\alpha \xi^\gamma - \Omega_{\alpha\gamma}\nabla_\beta \xi^\gamma \\ -\nabla_\alpha(\Omega_{\beta\gamma}\xi^\gamma) + \nabla_\beta(\Omega_{\alpha\gamma}\xi^\gamma) &= 0\end{aligned}$$

The first term of (32) gives the first three terms above, the second term of (32) the next three terms, and the last term of (32) the last two terms. But the expansion above holds (the third term cancels the fourth, the first, fifth, and seventh cancel, and the second, sixth, and eighth cancel). Hence (32) is true.

Since the curl of the symplectic field vanishes (Eqn. 11), (32) becomes

$$\pounds_\xi \Omega_{\alpha\beta} + 2\nabla_{[\alpha}(\Omega_{\beta]\gamma}\xi^\gamma) = 0 \tag{33}$$

Thus, ξ^α is an infinitesimal canonical transformation if and only if the curl of $\Omega_{\beta\gamma}\xi^\gamma$ vanishes. But (at least, locally) the curl of a covariant vector field on a manifold vanishes if and only if it is a gradient. Thus, ξ^α is an infinitesimal canonical transformation if and only if $\Omega_{\beta\gamma}\xi^\gamma = \nabla_\beta A$ for some scalar field A on phase space. Contracting this equation with $\Omega_{\beta\alpha}$, we find: ξ^α is an infinitesimal canonical transformation if and only if $\xi^\alpha = \Omega^{\beta\alpha}\nabla_\beta A$ for some A. This A is called the **generator** (or **generating function**) of the infinitesimal canonical transformation.

Thus, observables generate infinitesimal canonical transformations. It should also be clear why one often says that "the evolution of the system is the unfolding of a sequence of infinitesimal canonical transformations". (We would say: "The Hamiltonian vector field is an infinitesimal canonical transformation.")

Thus, infinitesimal canonical transformations can be described in terms of scalar fields on phase space rather than vector fields (an enormous simplification). What condition must an observable A satisfy in order that the infinitesimal canonical transformation it generates be an infinitesimal symmetry? This is easy to answer. We ask that the Lie derivative of H in the direction of $\Omega^{\beta\alpha}\nabla_\beta A$ vanish. That is to say, we ask that $(\Omega^{\beta\alpha}\nabla_\beta A)(\nabla_\alpha H) = 0$. That is to say, we ask that the Poisson bracket, $\{A, H\} = 0$. But, from (17), this is precisely the statement that A be constant along each dynamical trajectory, *i.e.*, that A be a constant of the motion. To summarize, we have shown that every constant of the motion generates an infinitesimal symmetry, and that, conversely, every symmetry is generated by a constant of the motion.

12. Algebraic Observables

Recall that the first step in the description of the system was the assignment to that system of a configuration space. This configuration space then led to phase space with its symplectic structure. Note, however, that essentially all of the notions of mechanics relate directly to phase space rather than configuration space. That is to say, the primary purpose in introducing configuration space has been to obtain phase space. In fact, one could just as well have treated mechanics as follows:

A system is described by a certain symplectic manifold, called phase space, on which there is a certain scalar field, called the Hamiltonian. One then has dynamical trajectories, Poisson brackets, canonical transformations, etc.

The present section is devoted to certain constructions which use, in an essential way, the presence of configuration space.

In sect. 10, we introduced configuration and momentum observables. We wish to generalize this notion. Consider a tensor field $R^{a_1 \cdots a_r}$, with r indices, on configuration space, and suppose that this tensor field is symmetric. (That is, suppose $R^{a_1 \cdots a_r} = R^{(a_1 \cdots a_r)}$, where round brackets, surrounding a collection of tensor indices, mean "add the expression to all expressions obtained by changing the order of the surrounded indices, and divide by the number of terms".) We associate, with this tensor field, an observable. Let (q, p_a) be a point of phase space. Then, $R^{a_1 \cdots a_r}(q)p_{a_1} \cdots p_{a_r}$ is a number. Thus, we have assigned a number to each point of phase space, *i.e.*, we have defined an observable. We shall write this observable $R^{a_1 \cdots a_r} p_{a_1} \cdots p_{a_r}$ or sometimes as $O(R)$. For $r = 0$ (*i.e.*, when $R^{a_1 \cdots a_r}$ has no indices), we obtain a configuration variable; for $r = 1$, a momentum observable. Note, however, that we also obtain observables which are quadratic, cubic, *etc.*, in the momenta.

Our goal is to express, in terms of the tensor fields, the results on various operations on these observables.

Evidently, one can add these totally symmetric tensor fields (*i.e.*, $R^{a_1 \cdots a_r}$ and $S^{a_1 \cdots a_s}$) when they have the same number of indices (*i.e.*, when $r = s$). There is also a product. If $R^{a_1 \cdots a_r}$ and $S^{a_1 \cdots a_s}$ are two such fields, we define their product, written $R \cap S$ (indices suppressed), by $R^{(a_1 \cdots a_r} S^{a_{r+1} \cdots a_{r+s})}$. Note that the number of indices in the product is $r + s$. Obviously, the product is also associative and commutative, and distributive over addition. Finally, we define a bracket relation on such tensor fields. Let $[R, S]$ (indices suppressed) be the

47

tensor field

$$rR^{ma_1\cdots a_{r-1}}\nabla_m S^{a_r\cdots a_{r+s-1}} - sS^{ma_1\cdots a_{s-1}}\nabla_m R^{a_s\cdots a_{r+s-1}} \qquad (34)$$

First note, from (6), that this expression is independent of the choice of derivative operator. Next, note that the bracket is antisymmetric, additive, and satisfies the Leibniz rule: $[R, S] = -[S, R]$, $[R + S, T] = [R, T] + [S, T]$, $[R \cap S, T] = R \cap [S, T] + [R, T] \cap S$. It is also true (and it can be verified directly from (34)) that this bracket relation satisfies the Jacobi identity: $[[R, S], T] + [[S, T], R] + [[T, R], S] = 0$. Finally, note that, when $r = 1$, (*i.e.*, when R is a vector), our bracket is just the Lie derivative $[R, S] = \pounds_R S$.

We now claim: the three operations (sum, product, bracket) introduced above on symmetric tensor fields on configuration space correspond precisely to the three operations (sum, product, Poisson bracket) on the corresponding observables. That is to say,

$$O(R + S) = O(R) + O(S) \qquad (35)$$

$$O(R \cap S) = O(R)O(S) \qquad (36)$$

$$O([R, S]) = \{O(R), O(S)\} \qquad (37)$$

We first note that (35) and (36) are obvious. To prove (37), one first checks, from (10), that it is valid when only configuration and momentum observables are permitted. But any tensor field can be written as a sum of outer products of vectors. Inserting such sums in (37), using the Leibniz rule (on the left, for the bracket; on the right, for the Poisson bracket) and the fact that (37) holds for vectors, we obtain (37) in general. This proves our assertion.

Thus, all the operations on observables yields algebraic observables (*i.e.*, those obtained from tensor fields on configuration space) when applied to algebraic observables. In fact, nearly every observable one normally deals with in mechanics is algebraic. An example follows.

Consider the Hamiltonian $H = g^{ab}p_a p_b + A^a p_a + V$, where g^{ab}, A^a, and V are fields on configuration space (with, of course, g^{ab} symmetric). The first term, for example, could be the Hamiltonian of a free particle in Euclidean space, where g^{ab} is the usual metric of Euclidean space. The first and third terms could be the Hamiltonian of a particle in a potential (V). The first and second terms could be the Hamiltonian of a charged particle (where A^a is a vector potential for the magnetic field). There appear, in fact, to be very few, if any, systems whose Hamiltonian cannot be put into this form. Let us ask when a momentum observable, $\xi^a p_a$, is a constant of motion for this Hamiltonian. The condition is that the Poisson bracket of the Hamiltonian and the observable vanish. Thus, we require that $\{\xi, g\} = 0$, $\{\xi, A\} = 0$, $\{\xi, V\} = 0$. That is to say, we require

$$\pounds_\xi g^{ab} = 0, \qquad \pounds_\xi A^b = 0, \qquad \pounds_\xi V = 0.$$

But this is just the statement that the "infinitesimal diffeomorphism" on configuration space generated by ξ^a leave invariant the various tensor fields (on configuration space) which go into the Hamiltonian.

Suppose our system is a free particle, so $A^a = 0$ and $V = 0$. Then, $\xi^a p_a$ is a constant of the motion whenever $\pounds_\xi g^{ab} = 0$, where g^{ab} is the metric of Euclidean space. There are six such vector fields on Euclidean space, representing translations and rotations. The constants of motion associated with translations are what are usually called *momenta*. The constants of motion associated with the rotations are what are usually called *angular momenta*. Now suppose we introduce a potential V, and ask which of our constants of motion remain constants of the motion. It is precisely those which also satisfy $\pounds_\xi V = \xi^a \nabla_a V = 0$. In other words, it is the symmetries of Euclidean space which also leave invariant the potential. If, for example, V is spherically symmetric, there will be three such vector fields, ξ_1^a, ξ_2^a, and ξ_3^a (namely, those which represent rotations about the origin of spherical symmetry of V). Set $J = \xi_1^a \cap \xi_1^a + \xi_2^a \cap \xi_2^a + \xi_3^a \cap \xi_3^a$, a symmetric, second-rank tensor field on configuration space. By the Jacobi identity, $O(J)$ is also a constant of the motion (a quadratic one). This constant of the motion is called the *squared angular momentum* (for a spherically symmetric system). The introduction of an A^a destroys, in general, these constants of the motion—unless, of course, the A^a happens to also be invariant under the corresponding infinitesimal diffeomorphism on configuration space.

Finally, note that constants of the motion remain constants of the motion under addition, product, and Poisson brackets. This is true, in particular, of algebraic constants of the motion.

Part III

Quantum Mechanics

13. Introduction

Mechanics is pretty, tidy, and natural looking. It suffers, however, from one serious flaw: Nature does not seem to behave that way. Specifically, quantum mechanics has been found to represent a more accurate description of Nature than classical mechanics. Roughly speaking, quantum mechanics is a "smeared out version of classical mechanics, where the amount of smearing is governed by a certain constant, Planck's constant $\hbar$." This last statement is, even as a "roughly speaking", somewhat unsatisfactory. It suggests that the principles of quantum mechanics are somehow to be imposed on a classically defined system, that quantum mechanics is some sort of thin veneer over classical mechanics. Thus, one often speaks of "quantizing" a classical system. (The situation is perhaps analogous to that in which one first treats matter as a continuum, and then "atomizes" it.) Systems, apparently, "really are" quantum systems—classical mechanics is merely a simplified approximation for the limit in which one is insensitive to the "smearing". Nonetheless, it is common, presumably because classical mechanics makes good contact with everyday life, to regard quantum mechanics as some sort of correction to classical mechanics.

Our goal is to quantize the system discussed in part II, *i.e.*, to write down the quantum description of various systems, and check that this description reduces, in an appropriate limit, to the classical description. It would perhaps be more logical, as the above remark suggests, to do part III before part II. We have artificially reversed the order to provide motivation: part II can be motivated from everyday life, and part III can be motivated from part II.

There are at least three essentially equivalent formulations of quantum mechanics: *Schrodinger, Heisenberg, and algebraic.* Essentially equivalent formulations, however, need not necessarily be of equal value: one may give better insight, be more readily adapted to other contexts, or suggest exotic generalizations, than another. The view is not uncommon that the Schrodinger formulation of quantum mechanics is less natural than the others. The reason is that the Schrodinger formulation seems to fit badly with the principle of relativity. Nonetheless, the Schrodinger formulation does tie in nicely with classical mechanics, and so we shall begin with this point of view.

Let us, for the moment, regard classical mechanics as consisting of states, observables, and dynamics. The states, of course, are the points of phase space, the observables scalar fields on phase space. We can regard the states and observables as the kinematics. The dynamics consists of the Hamiltonian dynamical trajectories, *etc.*

Each of these three classical notions has an analog in quantum mechanics. The space of states in classical mechanics is to be replaced by a corresponding (although different) space of states in quantum mechanics. In either case, the system is to be regarded as possessing, at each instant of time, one of its possible states. Dynamics plays essentially the same role in quantum mechanics as in classical mechanics: it is the means by which one describes the successive states through which a system passes with time. It is with the observables, however, that a significant departure between classical and quantum mechanics, apparently, takes place. Classically, an observable assigns a number to each state—a feature one might expect to be characteristic of what one means by a "measuring instrument". It might seem reasonable, therefore, to suppose that observables also in quantum mechanics will have this same characteristic property. However, observables emerge initially in quantum mechanics as mappings from states to states. This mathematical feature is presumably a reflection of the following physical fact: *observations on a classical system can be performed with negligible effect on the system observed, while this is not the case for a quantum system.* In other words, the mere assignment of a number to each state is not enough information to define a quantum observable; one must also give information regarding the state into which the system will be thrown by this act of observation. This "extra information" is carried by having observables map states to states rather than to real numbers.

14. Densities. Integrals

The things which can be integrated over a manifold, it turns out, are, not scalar fields, but things called *scalar densities*. Scalar densities, in turn, represent a special case of *tensor densities*. We introduce here the appropriate definitions and properties.

Let M be an n-dimensional manifold. An **alternating tensor**, $\epsilon^{a_1 \cdots a_n}$, at a point of M is an n-index contravariant tensor at that point which is antisymmetric, *i.e.*, which reverses sign under interchange of any two of its indices, *i.e.*, which satisfies $\epsilon^{[a_1 \cdots a_n]} = \epsilon^{a_1 \cdots a_n}$. If ϵ and ϵ' are two alternating tensors at the same point, with the second nonzero, then there exists a number α such that $\epsilon^{a_1 \cdots a_n} = \alpha \epsilon'^{a_1 \cdots a_n}$. (Proof: Choose a basis $v_{1a}, \ldots, v_{na}$, for the covariant vectors at the point, and note that, by antisymmetry, an alternating tensor is completely and uniquely determined by the value of the number $\epsilon^{a_1 \cdots a_n} v_{1a_1} \cdots v_{na_n}$.) Thus, the alternating tensors at a point of a manifold form a one-dimensional vector space. An **alternating tensor field** is an n-index, contravariant, antisymmetric tensor field on M, *i.e.*, an assignment (smoothly) of an alternating tensor to each point of M.

Fix a real number s. A **tensor density**, *e.g.*, $T^p{}_{bc}{}^w$, at a point of M is a mapping from alternating tensors $\epsilon^{a_1 \cdots a_n}$ at that point to tensors $T^p{}_{bc}{}^w(\epsilon)$ at that point such that

$$T^p{}_{bc}{}^w(\alpha\epsilon) = (\alpha)^s T^p{}_{bc}{}^w(\epsilon) \tag{38}$$

for any real number α. The number s is called the **weight** of the tensor density. Since any two alternating tensors are proportional, (38) implies that a knowledge of $T^p{}_{bc}{}^w(\epsilon)$ for one nonzero alternating tensor $\epsilon^{a_1 \cdots a_n}$ determines $T^p{}_{bc}{}^w(\epsilon)$ for all ϵ, and hence, determines the tensor density $T^p{}_{bc}{}^w$. Thus, a choice of alternating tensor at a point defines a one-to-one correspondence between tensor densities at that point and tensors at that point. Roughly speaking, a tensor density is a tensor having certain "scaling behavior" with alternating tensors. Note, from (38), that a tensor density of weight zero is independent of the alternating tensor. That is to say, tensor densities of weight zero are just ordinary tensors.

Example: Let $\kappa_{a_1 \cdots a_n}$ be a tensor at a point of M. For any alternating tensor $\epsilon^{a_1 \cdots a_n}$ at this point, set $\phi(\epsilon) = \kappa_{a_1 \cdots a_n} \epsilon^{a_1 \cdots a_n}$. Thus, we have defined a mapping from alternating tensors at a point to scalars at that point (*i.e.*, to real numbers). It is immediate that this mapping satisfies (38) with $s = 1$. Hence, $\phi(\epsilon)$ is a scalar density of weight $+1$. More generally, since tensor densities of weight $+1$ define, via (38), linear mappings from alternating tensors to tensors, every tensor density of weight $+1$ can be represented, as above, by a tensor. Similarly, $\lambda_{a_1\cdots a_n b_1\cdots b_n}\epsilon^{a_1\cdots a_n}\epsilon^{b_1\cdots b_n}$ is a scalar density of weight $+2$. It is now clear that every tensor density of positive integral weight can be represented as a tensor. Now let $\eta^{a_1\cdots a_n}$ be a nonzero alternating tensor at a point of M. Then, for $\epsilon^{a_1\cdots a_n}$ another alternating tensor at that point, set $\chi(\epsilon) = \alpha$, where α is the number such that $\eta^{a_1\cdots a_n} = \alpha\epsilon^{a_1\cdots a_n}$. Clearly, this $\chi(\epsilon)$ (a mapping from alternating tensors to scalars) is a scalar density of weight -1. Thus, a nonzero alternating tensor defines a scalar density of weight -1. Similarly, every tensor density of negative integral weight can be represented by a tensor. The purpose of introducing tensor densities is to enable us to deal with non-integral weights.

A **tensor density field** on M is a (smooth) assignment of a tensor density to each point of M, where the weight is the same for each point of M. (Here, of course, "smooth" means that a smooth tensor field results from a smooth alternating tensor field.)

Which of our tensor operations are applicable to tensor density fields? Let $T^a{}_{cd}$ and $T'^a{}_{cd}$ be tensor density fields, and, for any alternating tensor ϵ, set $(T^a{}_{cd} + T'^a{}_{cd})(\epsilon) = T^a{}_{cd}(\epsilon) + T'^a{}_{cd}(\epsilon)$. Now let T and T' have weights s and s', respectively. Then, clearly, the right hand side of this equation is a tensor density (*i.e.*, the right hand side satisfies (38)) only when $s = s'$. Thus, the sum of two tensor density fields is defined precisely when they have the same weight (and, of course, the same index structure). The weight of the sum is the weight of each addend. Similarly, (*e.g.*, $(T^a{}_{cd}W_{pq})(\epsilon) = T^a{}_{cd}(\epsilon)W_{pq}(\epsilon)$), the outer product of any two tensor density fields is defined. The weight of the product is the sum of the weights of the factors. (Note that, for weight zero, *i.e.*, for ordinary tensor fields, these operations reduce to ordinary sum and outer product.) Finally, index substitution and contraction are defined in the obvious way for tensor densities, and, when applied, yield a density of the same weight. In short, except for the provision that addition is defined only between densities having the same weight, all operations on tensors extend to operations on tensor densities.

Fix a derivative operator ∇_a on M, so this ∇_a acts on tensor fields. We wish to extend this action to tensor densities. Let $\epsilon^{a_1\cdots a_n}$ be an alternating tensor field on M. Then $\nabla_a\epsilon^{a_1\cdots a_n} = \lambda_a\epsilon^{a_1\cdots a_n}$ for some covariant vector field λ_a. (Proof: the left hand side of this equation is antisymmetric in $a_1,\cdots,a_n$. But two alternating tensors are proportional.) We now define the derivative of a tensor density, *e.g.*, $T^b{}_{cd}$, by

$$(\nabla_a T^b{}_{cd})(\epsilon) = \nabla_a(T^b{}_{cd}(\epsilon)) - s\lambda_a T^b{}_{cd}(\epsilon) \tag{39}$$

where s is the weight of $T^b{}_{cd}$. That is to say, $\nabla_a T^b{}_{cd}$ is to be the tensor density field whose action on any alternating tensor field ϵ is given by the right hand side of (39). We must check that the right hand side of (39) indeed satisfies (38). To this end, let $\epsilon'^{a_1\cdots a_n} = \alpha\epsilon^{a_1\cdots a_n}$. Then $\nabla_a(\epsilon'^{a_1\cdots a_n}) = \nabla_a(\alpha\epsilon^{a_1\cdots a_n}) = \alpha\nabla_a(\epsilon^{a_1\cdots a_n}) + (\nabla_a\alpha)\epsilon^{a_1\cdots a_n}$. Hence, $\lambda'_a = \lambda_a + \alpha^{-1}(\nabla_a\alpha)$. We then have, for the right side of (39),

$$\nabla_a(T^b{}_{cd}(\alpha\epsilon)) - \lambda'_a T^b{}_{cd}(\lambda\epsilon)$$
$$= \nabla_a(\alpha^s T^b{}_{cd}(\epsilon)) - (\lambda_a + \alpha^{-1}\nabla_a\alpha)\alpha^s T^b{}_{cd}(\epsilon)$$
$$= \alpha^s \nabla_a T^b{}_{cd}(\epsilon) + s\alpha^{s-1}(\nabla_a\alpha)T^b{}_{cd}(\epsilon)$$
$$- \lambda_a \alpha^s T^b{}_{cd}(\epsilon) - \alpha^{s-1}(\nabla_a\alpha)T^b{}_{cd}(\epsilon)$$
$$= \alpha^s[\nabla_a(T^b{}_{cd}(\epsilon)) - \lambda_a T^b{}_{cd}(\epsilon)]$$

Thus, (39) indeed defines a tensor density field of weight s. To summarize, a derivative operator on tensor fields extends, via (39), to an operator on tensor density fields, where taking the derivative does not change the weight. Note, from (39), that a derivative operator on tensor density fields is additive and satisfies the Leibniz rule.

Now suppose that we have two derivative operators, ∇_a and ∇'_a, on M, related via (6). We wish to find the expression, analogous to (6), relating the actions of these operators on tensor densities. Let $\epsilon^{a_1\cdots a_n}$ be an alternating tensor field on M. Then, from (6), $\nabla'_a(\epsilon^{a_1\cdots a_n}) = \nabla_a(\epsilon^{a_1\cdots a_n}) - C^m{}_{am}\epsilon^{a_1\cdots a_n}$. That is to say, $\lambda'_a = \lambda_a - C^m{}_{am}$. Then, (39) implies $(\nabla'_a T^b{}_{cd} - \nabla_a T^b{}_{cd})(\epsilon) = (\nabla'_a - \nabla_a)T^b{}_{cd}(\epsilon) + sC^m{}_{am}T^b{}_{cd}(\epsilon)$. Since, using (6), we know how $(\nabla'_a - \nabla_a)$ acts on a tensor $T^b{}_{cd}(\epsilon)$, we can expand the first term on the right hand side to obtain $(\nabla'_a T^b{}_{cd} - \nabla_a T^b{}_{cd})(\epsilon) = -C^b{}_{am}T^m{}_{cd}(\epsilon) + C^m{}_{ac}T^b{}_{md}(\epsilon) + C^m{}_{ad}T^b{}_{cm}(\epsilon) + sC^m{}_{am}T^b{}_{cd}(\epsilon)$. Thus, the generalization of (6) is

$$(\nabla'_a T^b{}_{cd} - \nabla_a T^b{}_{cd}) = -C^b{}_{am}T^m{}_{cd} + C^m{}_{ac}T^b{}_{md}$$
$$+C^m{}_{ad}T^b{}_{cm} + sC^m{}_{am}T^b{}_{cd} \tag{40}$$

In short, things about tensor fields carry over with appropriate (and minor) modifications to things about tensor density fields.

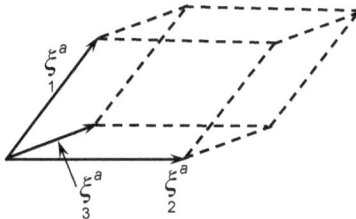

Our final task is to introduce integrals. At a point of the n-dimensional manifold M, choose N contravariant vectors, $\xi_1^a, \ldots, \xi_n^a$. Since a contravariant vector can be regarded as defining an "infinitesimal motion at a point", these n vectors can be regarded as defining an infinitesimal parallelepiped. We would like to associate with this figure a "volume". If we were to multiply any of the vectors $\xi_1^a, \ldots, \xi_n^a$ by a factor, the volume should be multiplied by that factor. If we add to any of these vectors a multiple of another vector, the volume should remain the same. Thus, the volume of this infinitesimal figure should be proportional to the alternating tensor $\xi_1^{[a_1}\cdots\xi_n^{a_n]}$. We can think of a volume element as assigning to each infinitesimal parallelepiped, i.e., to each alternating tensor, a number, linearly in the alternating tensor. In other words, a volume element should be a scalar density of weight $+1$. It is a scalar density of weight $+1$ which can be integrated over a manifold to obtain a number.

To actually define an integral over a manifold, we must go back to charts. Let ϕ be a scalar density of weight $+1$. Choose a chart, with coordinates $x^1, \ldots, x^n$. Consider the unique alternating tensor satisfying $\epsilon^{a_1 \cdots a_n}(\nabla_{a_1} x^1) \cdots (\nabla_{a_n} x^n) = 1$. Then the integral of ϕ over the region covered by this chart is, by definition, the integral $\int \phi(\epsilon) dx^1 \cdots dx^1$. It is because our original density was of weight $+1$ that this integral is independent of the choice of chart.

15. States

In classical mechanics, the states are points of phase space. In quantum mechanics, the **states** are complex-valued densities ψ on configuration space C, of weight $+\frac{1}{2}$, for which the integral

$$\int_C \bar{\psi}\psi = \|\psi\|^2 \tag{41}$$

converges (*i.e.*, is finite), where a bar denotes the complex conjugate. There is a sense in which a quantum state can be regarded as a "smeared out" classical state. The "configuration half", q, of a classical state (q, p_a), is represented by $\bar{\psi}\psi$, a density of weight $+1$. The density could, for example, be "peaked" near a point q of configuration space. The "lack of peaking" corresponds to the "amount of smearing out". The momentum information in a classical state is represented, within a quantum state, by the phase of the density ψ. The "direction in which the phase is changing fastest" replaces the "direction of p_a", while the rate of change of the phase represents the "magnitude of p_a". Thus, the two "point pieces of information", q and p_a, which go into a classical state are reflected, in a smeared out way, by the real and imaginary parts of the density ψ.

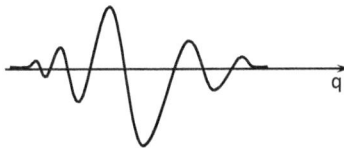

There is one remarkable thing that has happened in the transition from classical to quantum states. Recall that, in classical mechanics, the introduction of configuration space was essentially for motivational reasons. One could as well have carried out classical mechanics on phase space (a symplectic manifold) on which a scalar field, the Hamiltonian, was specified. On the other hand, right at the beginning in quantum mechanics—already in the specification of the states—configuration space plays an essential role. Suppose, for example, that we produced two manifolds whose cotangent bundles were identical as symplectic manifolds. Let this symplectic manifold be the phase space for some system. Then either of our two original manifolds could be regarded as the configuration space of the system. In classical mechanics, this choice makes no difference, for classical mechanics takes place on phase space. But, in quantum mechanics, the difference is crucial, for the space of states already depends essentially on the choice of configuration space. Stated more carefully, one might say that the physical world apparently attaches significance to configuration space itself, but that this significance seems to get lost in the passage to the classical limit.

Essentially, the only structure available on the space of classical states are the smoothness and symplectic structures. What structure is available on the space of quantum states? As it turns out, a great deal more. Firstly, we can add states, if ψ and ψ' are complex-valued densities on C of weight $+\frac{1}{2}$ for which the integral (41) converges, then $\psi+\psi'$ (sum of densities) is another. Furthermore, if ψ is a quantum state, and κ a complex number, then $\kappa\psi$ is another state. Thus, we can add states and multiply them with complex numbers. It is obvious that the space of quantum states thus becomes a complex vector space. Still more structure comes from (40). Given two states, ψ and ψ', we define their **inner product**, a complex number, by

$$\langle\, \psi' \,|\, \psi \,\rangle = \int_C \bar{\psi}'\psi \tag{42}$$

It is clear that the inner product satisfies the following properties:

1. $\langle\, \psi \,|\, \psi' + \kappa\psi'' \,\rangle = \langle\, \psi \,|\, \psi' \,\rangle + \kappa\langle\, \psi \,|\, \psi'' \,\rangle$

2. $\overline{\langle\, \psi \,|\, \psi' \,\rangle} = \langle\, \psi' \,|\, \psi \,\rangle$

3. $\langle\, \psi \,|\, \psi \,\rangle \geq 0$, equality holding when and only when $\psi = 0$

More generally, an **inner product space** is a complex vector space on which an inner product is defined, satisfying conditions (1), (2) and (3) above. Thus, the space of quantum states is an inner product space. An inner product space which is *complete* (in an appropriate sense) is called a **Hilbert space**. The inner product spaces which occur in quantum mechanics are almost never complete. However, there exists a procedure for "completing" an inner product space to obtain a Hilbert space. Thus, one sometimes regards the states of a quantum system as residing in a Hilbert space (*i.e.*, by suitably completing our space of states above). It makes essentially no difference in practice whether one choose to work in the inner product or the Hilbert space. Primarily to avoid having to introduce the technical construction of completing, we shall remain in the inner product space.

It is natural to ask why the very rich structure of the states of an actual system (*i.e.*, the quantum states) become lost in the passage to the classical limit. This is not at all clear (to me) except for the remark that, for some reason, *superposition is lost in the classical limit*.

16. Observables

We now introduce observables in quantum mechanics. It is convenient to begin summarizing some mathematics.

Let $\mathcal{H}$ be an inner product space. A mapping $O : \mathcal{H} \to \mathcal{H}$ which is linear (*i.e.*, which satisfies $O(\psi + \kappa\psi') = O(\psi) + \kappa O(\psi')$ for all elements ψ, ψ' of $\mathcal{H}$ and complex numbers κ) is called an **operator** on $\mathcal{H}$. Actually, it is convenient to slightly broaden this definition. A subspace $\mathcal{D}$ of $\mathcal{H}$ will be said to be **dense** if, given any element ψ of $\mathcal{H}$ and any number $\epsilon > 0$, there is an element ψ' of $\mathcal{D}$ such that $\|\psi - \psi'\| \leq \epsilon$. (Intuitively, a dense subset possesses elements arbitrarily close to any element of $\mathcal{H}$.) We shall also call linear mappings from a dense subspace $\mathcal{D}$ of $\mathcal{H}$ to $\mathcal{H}$ **operators**. (This is a technical point, necessitated by the fact that most of the operators we shall need are only defined, in any reasonable way, on a dense subspace of an inner product space.) An operator will be called **self-adjoint** if, whenever both ψ and ψ' are in the dense subspace on which the operator is defined, $\langle\, \psi' \,|\, O\psi \,\rangle = \langle\, O\psi' \,|\, \psi \,\rangle$.

An **observable** in quantum mechanics is a self-adjoint operator on the inner product space of states.

Of particular interest are observables which have a direct connection with classical observables. Ideally, one would like to formulate a rule for passing from any classical observable (a scalar field on phase space) to a corresponding quantum observable. One has, however, no reason to expect that such a rule will exist—and, apparently, none does. In fact, it is difficult to think, offhand, of any way whatever to pass from a scalar field on phase space to an operator on the space of certain densities on configuration space. From a purely mathematical point of view, it is perhaps surprising that any classical observables have quantum analogues. Perhaps one should ask for less—ask only that the algebraic observables go over to quantum mechanics. But, at this stage, even this appears to be impossible. It turns out that the classical observables which do have simple and unambiguous quantum versions are the configuration and momentum observables.

Let α be a scalar field on configuration space. Then, as we have seen, α may also be regarded as a scalar field on phase space, *i.e.*, as a classical observable. We now define a corresponding quantum observable: $Q(\alpha)\psi = \alpha\psi$. That is to say, $Q(\alpha)$ is the operator on the inner product space of states which, acting on a typical state ψ yields the state obtained by multiplying ψ (a density on C of weight $+\frac{1}{2}$) by α (a density on C of weight 0). The result is another state, so $Q(\alpha)$ is a mapping from states to states. These operators $Q(\alpha)$ will be called **configuration observables**: they are directly analogous to configuration observables in classical mechanics. Note that each $Q(\alpha)$ is self-adjoint: $\langle \psi' \,|\, Q(\alpha)\psi \rangle = \int \bar{\psi}'(\alpha\psi) = \int (\alpha\bar{\psi}')\psi = \langle Q(\alpha)\psi' \,|\, \psi \rangle.$.

The momentum observables are slightly more subtle. Let ξ^a by a vector field on configuration space, so $\xi^a p_a$ is a classical momentum observable. We wish to write down a corresponding quantum observable. The standard rule for this transition is that of "replacing the classical momentum p_a by $(\frac{\hbar}{i})\nabla_a$". Thus, one might expect that the appropriate quantum operator is that whose action on a state ψ yields $(\frac{\hbar}{i})\xi^a\nabla_a\psi$. However, there are two difficulties with this choice. Firstly, the operator is not self-adjoint. Indeed,

$$\int \bar{\psi}'((\tfrac{\hbar}{i})\xi^a\nabla_a\psi) - \int ((\tfrac{\hbar}{i})\xi^{\bar{a}}\nabla_a\bar{\psi}')\psi$$

$$= (\tfrac{\hbar}{i}) \int (\bar{\psi}'\xi^a\nabla_a\psi + (\xi^a\nabla_a\bar{\psi}')\psi)$$

$$= (\tfrac{\hbar}{i}) \int (\nabla_a(\bar{\psi}'\psi\xi^a) - \bar{\psi}'\psi\nabla_a\xi^a)$$

$$= -(\tfrac{\hbar}{i}) \int \bar{\psi}'\psi\nabla_a\xi^a$$

But the right hand side is not in general zero. Even more serious is the fact that the expression $(\frac{\hbar}{i})\xi^a\nabla_a\psi$ is not independent of the choice of derivative operator. In fact, we have from (40)

$$\xi^a\nabla'_a\psi - \xi^a\nabla_a\psi = \frac{1}{2}\xi^a C^m{}_{am}\psi \neq 0.$$

It is a remarkable fact that a single change in our operator suffices to correct both of these difficulties. For ξ^a, a vector field on configuration space, we define an operator, $P(\xi)$, with the following action on a typical state ψ:

$$P(\xi)\psi = (\tfrac{\hbar}{i})(\xi^a\nabla_a\psi + \tfrac{1}{2}\psi\nabla_a\xi^a) \tag{43}$$

It is immediate from the above that the second term on the right in (43) makes this operator self-adjoint, and, at the same time, produces an operator independent of the choice of derivative operator. (This last property is vital, for we have, on configuration space, no preferred derivative operator.) The quantum observables $P(\xi)$ will be called **momentum observables**. (Note that the usual angular momentum operators in quantum mechanics are also, by our definition, momentum observables.) The quantum momentum observables are, of course, analogous to the classical momentum observables.

There are three fundamental sets of equations satisfied by these configuration and momentum observables. The first, and easiest, are linearity in the argument:

$$Q(\alpha + \beta) = Q(\alpha) + Q(\beta) \tag{44}$$

$$P(\xi + \eta) = P(\xi) + P(\eta) \tag{45}$$

These properties are immediate from the definitions. The second set of equations are two anti-commutator relations. For A and B operators, we define the **anti-commutator** by $\{A, B\} = AB + BA$. We have, immediately from the definition, that $Q(\alpha)Q(\beta) = Q(\alpha\beta)$. Hence,

$$\{Q(\alpha), Q(\beta)\} = 2Q(\alpha\beta) \tag{46}$$

The anti-commutator between configuration and momentum observables also takes a simple form. For ψ a state, we have

$$
\begin{aligned}
&\{P(\xi), Q(\alpha)\}\psi = P(\xi)\left(Q(\alpha)\psi\right) + Q(\alpha)\left(P(\xi)\psi\right) \\
&= P(\xi)(\alpha\psi) + \alpha(\tfrac{\hbar}{i})(\xi^a \nabla_a \psi + \tfrac{1}{2}\psi\nabla_a \xi^a) \\
&= (\tfrac{\hbar}{i})\left((\xi^a \nabla_a(\alpha\psi) + \tfrac{1}{2}(\alpha\psi)\nabla_a\xi^a) + \alpha(\xi^a \nabla_a\psi + \tfrac{1}{2}\psi\nabla_a\xi^a)\right) \\
&= (\tfrac{\hbar}{i})\left(\xi^a(\nabla_a\alpha)\psi + 2\alpha\xi^a\nabla_a\psi + \alpha\psi\nabla_a\xi^a\right) \\
&= (\tfrac{\hbar}{i})\left(2\alpha\xi^a\nabla_a\psi + \psi\nabla_a(\alpha\xi^a)\right) \\
&= 2(\tfrac{\hbar}{i})\left(\alpha\xi^a\nabla_a\psi + \tfrac{1}{2}\psi\nabla_a(\alpha\xi^a)\right) \\
&= 2P(\alpha\xi)\psi
\end{aligned}
$$

Hence, since ψ is arbitrary,

$$\{P(\xi), Q(\alpha)\} = 2P(\alpha\xi) \tag{47}$$

Finally, we derive the three (canonical) commutation relations. For A and B operators, we set $[A, B] = AB - BA$. It is immediate from the definition that

$$[Q(\alpha), Q(\beta)] = 0 \tag{48}$$

Furthermore, we have

$$
\begin{aligned}
&[P(\xi), Q(\alpha)]\psi = P(\xi)\left(Q(\alpha)\psi\right) - Q(\alpha)\left(P(\xi)\psi\right) \\
&= P(\xi)(\alpha\psi) - \alpha(\tfrac{\hbar}{i})(\xi^a \nabla_a\psi + \tfrac{1}{2}\psi\nabla_a\xi^a) \\
&= (\tfrac{\hbar}{i})\left((\xi^a\nabla_a(\alpha\psi) + \tfrac{1}{2}(\alpha\psi)\nabla_a\xi^a) - \alpha(\xi^a\nabla_a\psi + \tfrac{1}{2}\psi\nabla_a\xi^a)\right) \\
&= (\tfrac{\hbar}{i})\left(\xi^a(\nabla_a\alpha)\psi\right) \\
&= (\tfrac{\hbar}{i})Q(\xi^a\nabla_a\alpha)\psi
\end{aligned}
$$

whence

$$[P(\xi), Q(\alpha)] = (\tfrac{\hbar}{i})Q(\xi^a\nabla_a\alpha) \tag{49}$$

Finally, we obtain the commutation relations for the momentum observables:

$$[P(\xi), P(\eta)]\psi$$
$$= (\tfrac{\hbar}{i})^2[\xi^a\nabla_a(\eta^b\nabla_b\psi + \tfrac{1}{2}\psi\nabla_b\eta^b) + (\eta^b\nabla_b\psi + \tfrac{1}{2}\psi\nabla_b\eta^b)\nabla_a\xi^a$$
$$- \eta^a\nabla_a(\xi^b\nabla_b\psi + \tfrac{1}{2}\psi\nabla_b\xi^b) - \tfrac{1}{2}(\xi^b\nabla_b\psi + \tfrac{1}{2}\psi\nabla_b\xi^b)\nabla_a\eta^a]$$
$$= (\tfrac{\hbar}{i})^2[(\xi^a\nabla_a\eta^b - \eta^a\nabla_a\xi^b)\nabla_b\psi + (\xi^a\eta^b - \eta^a\xi^b)\nabla_a\nabla_b\psi$$
$$+ \tfrac{1}{2}\psi[\xi^a\nabla_a\nabla_b\eta^b - \eta^a\nabla_a\nabla_b\xi^b]$$
$$= (\tfrac{\hbar}{i})^2[(\xi^a\nabla_a\eta^b - \eta^a\nabla_a\xi^b)\nabla_b\psi + \tfrac{1}{2}\psi\nabla_b(\xi^a\nabla_a\eta^b - \eta^a\nabla_a\xi^b)$$
$$+ (\xi^a\eta^b - \eta^a\xi^b)\nabla_a\nabla_b\psi + \psi(\xi^a\nabla_{[a}\nabla_{b]}\eta^b - \eta^a\nabla_{[a}\nabla_{b]}\xi^b)]$$
$$= (\tfrac{\hbar}{i})P(\mathcal{L}_\xi\eta)\psi + (\tfrac{\hbar}{i})^2[(\xi^a\eta^b - \eta^a\xi^b)\nabla_a\nabla_b\psi$$
$$+ \psi(\xi^a\nabla_{[a}\nabla_{b]}\eta^b - \eta^a\nabla_{[a}\nabla_{b]}\xi^b)]$$
$$= (\tfrac{\hbar}{i})P(\mathcal{L}_\xi\eta)\psi + (\tfrac{\hbar}{i})^2\psi^{-1}\nabla_a\nabla_b(\psi^2\xi^{[a}\eta^{b]})$$

The final result

$$[P(\xi), P(\eta)] = (\tfrac{\hbar}{i})P(\mathcal{L}_\xi\eta) \tag{50}$$

will follow from the calculation above if we show that the last term on the far right side [i.e., $(\tfrac{\hbar}{i})^2\psi^{-1}\nabla_a\nabla_b(\psi^2\xi^{[a}\eta^{b]})$] vanishes. Setting $F^{ab} = F^{[ab]} = \psi^2\xi^{[a}\eta^{b]}$, we have, for an arbitrary scalar field α

$$\int \alpha\nabla_a\nabla_b F^{ab} = -\int (\nabla_a\alpha)\nabla_b F^{ab} = \int F^{ab}\nabla_a\nabla_b\alpha$$
$$= \int F^{ab}\nabla_{[a}\nabla_{b]}\alpha = 0$$

where we have integrated twice by parts. Hence, $\nabla_a\nabla_b F^{ab} = 0$, and our result, (50), follows.

Eqns. (48), (49), (50) are called the **canonical commutation relations**. Note that, in each case, the commutator of the observables is $(\tfrac{\hbar}{i})$ times the Poisson bracket of the corresponding classical observables. It is interesting that the quantum observables manage to encompass, in the commutation relations, two features of the classical observables. Firstly, the commutators all contain a factor of $\hbar$ on the right. Thus, in the classical limit ($\hbar \to 0$), the observables commute, just as do classical observables (where product is outer product). On the other hand, the Poisson bracket of classical observables is also reflected in the commutators.

The fundamental properties of the configuration and momentum observables are given by Eqns. (44)-(50). These operators and properties are incorporated, in a somewhat peculiar way, into the standard formulation of Schrodinger quantum mechanics. Consider, for example, a free particle. One introduces configuration observables Q_x, Q_y, and Q_z, obtained from the scalar fields x, y, and z on Euclidean space (configuration space). These commute with each other. From (44) and (46), configuration observables associated with any scalar field which is a polynomial in x, y, z can be constructed as combinations of the observables Q_x, Q_y, and Q_z. One introduces momentum observables P_x, P_y, and

P_z, defined by unit vector fields in the x, y, and z directions, respectively. Using (45) and (47), the momentum observable associated with any vector field in Euclidean space whose Euclidean components are polynomials in x, y, and z can be constructed from Q_x, Q_y, Q_z, P_x, P_y, P_z. Thus, using (44)-(47), one recovers essentially (*i.e.*, insofar as polynomials are "essentially all functions") all of the configuration and momentum observables in terms of just six. On these six observables, the canonical commutation relations are imposed. These, of course, are just (48)-(50) in the special cases. Since all other observables are expressed in terms of these six, the commutation relations on the six suffice to obtain all commutation relations. Our formulae (48)-(50) are just these "general commutation relations for all configuration and momentum observables". In short, instead of many observables, subject to (44)-(50), one uses (44)-(47) to reduce to six observables, on which (48)-(50) are imposed. The advantage of our formulation, in my view, is that one avoids making special choices (of which six observables are to be given special status), and treats everything on a simple, one-shot basis. For examples, the angular momentum commutators for the free particle are special cases of (50).

17. Higher Order Observables

The introduction of configuration and momentum observables works out remarkably well. The definitions are simple and natural, and the resulting operators satisfy every condition one could reasonably ask for. One is led by this success to try to carry over into quantum mechanics classical observables which are quadratic, cubic, *etc.* in momentum. This we attempt, unsuccessfully, in the present section.

First note that it is easy to construct observables which involve second, third, *etc.* derivatives of ψ. For example, let ξ^a be a vector field on configuration space. Then $P(\xi)P(\xi)$ is an observable whose explicit action, from (43), is

$$P(\xi)P(\xi)\psi = (\tfrac{\hbar}{i})^2 \left[\xi^a \xi^b \nabla_a \nabla_b \psi + (\xi^a \nabla_b \xi^b + \xi^b \nabla_b \xi^a)\nabla_a \psi \right.$$
$$\left. + (\tfrac{1}{2}\xi^a \nabla_a(\nabla_b \xi^b) + \tfrac{1}{4}(\nabla_b \xi^b)^2)\psi \right]$$

More generally, any anti-commutator of observables is an observable. It would be natural to regard $\{P(\xi), P(\eta)\}$ as the quantum observable $T^{ab}p_a p_b$, where $T^{ab} = \xi^{(a}\eta^{b)}$. (That is to say, we suitably generalize (46) and (47).) Unfortunately, a problem arises. Let λ be any scalar field on configuration space C which vanishes nowhere (so λ^{-1} exists). Then it is also true that $T^{ab} = (\lambda^{-1}\xi^{(a})(\lambda\eta^{b)})$. However, using (46)-(50), it is easily checked in general $\{P(\lambda^{-1}\xi), P(\lambda\eta)\} \neq \{P(\xi), P(\eta)\}$. More generally, suppose we are given a quadratic classical observable, $T^{ab}p_a p_b$. One can always write T^{ab} as a sum of symmetrized outer products of vector fields. One would like to call the corresponding sum of anti-commutators of momentum observables the quantum version of the classical observable $T^{ab}p_a p_b$. However, as illustrated above, the resulting quantum observable will in general depend on the decomposition chosen for T^{ab}. In short, we fail, in this way, to associate with classical observables which are quadratic, cubic, *etc.* in momentum corresponding (unique) quantum observables.

Let $T^{ab}p_a p_b$ be a classical observable. We proceed, in a more direct way, to try to associate with this a quantum observable. Clearly, the most general candidate is

$$\alpha T^{ab}\nabla_a \nabla_b \psi + \beta(\nabla_b T^{ab})\nabla_a \psi + \gamma(\nabla_a \nabla_b T^{ab})\psi \qquad (51)$$

where α, β, and γ are real numbers. In order to be acceptable, this candidate must, at least, satisfy two conditions: it must be self-adjoint, and it must be independent of the choice of derivative operator.

We ask what α, β, and γ must be in order that each of these conditions, in turn, be satisfied. Evidently, we have

$$\int \bar{\psi}'(\alpha T^{ab}\nabla_a\nabla_b\psi + \beta\nabla_b T^{ab}\nabla_a\psi + \gamma\nabla_a\nabla_b T^{ab}\psi)$$

$$= \int (-\alpha\nabla_b\psi\nabla_a(T^{ab}\bar{\psi}') + \beta\bar{\psi}'\nabla_b T^{ab}\nabla_a\psi + \gamma\nabla_a\nabla_b T^{ab}\psi\bar{\psi}')$$

$$= \int (-\alpha T^{ab}(\nabla_a\bar{\psi}')(\nabla_b\psi) + (\beta - \alpha)\bar{\psi}'\nabla_b T^{ab}\nabla_a\psi + \gamma\psi\bar{\psi}'\nabla_a\nabla_b T^{ab})$$

where, in the second step, we have integrated by parts, and, in the third step, we used the symmetry of T^{ab}. Hence, a necessary and sufficient condition that T be self-adjoint is that $\alpha = \beta$. To investigate behavior under change in the choice of derivative operator, we use (40):

$$\alpha T^{ab}\nabla'_a\nabla'_b\psi + \beta(\nabla'_a T^{ab})(\nabla'_b\psi) + \gamma(\nabla'_a\nabla'_b T^{ab})\psi$$

$$= \alpha T^{ab}\nabla'_a(\nabla_b\psi + \tfrac{1}{2}C^m{}_{bm}\psi)$$

$$+\beta(\nabla_a T^{ab} - C^b{}_{am}T^{mn} - C^m{}_{am}T^{ab})(\nabla_b\psi + \tfrac{1}{2}C^p{}_{bp}\psi)$$

$$+\gamma\psi\nabla'_a(\nabla_b T^{ab} - C^a{}_{mn}T^{mn} - C^m{}_{bm}T^{ab})$$

$$= \alpha T^{ab}\nabla_a\nabla_b\psi + \beta(\nabla_a T^{ab})\nabla_b\psi + \gamma(\nabla_a\nabla_b T^{ab})\psi$$

$$+(\nabla_a\psi)((\tfrac{3}{2}\alpha - \beta)T^{ab}C^m{}_{bm} + (\alpha - \beta)T^{mn}C^a{}_{mn})$$

$$+\psi(\ldots)$$

where the dots indicate a rather complicated expression which involves (among other things) γ. It is immediate from this expression that our operator is independent of the choice of derivative operator precisely when $\alpha = \beta = \gamma = 0$, $i.e.$, when and only when we have the zero operator. This failure is, of course, just a more sophisticated version of our earlier failure.

Thus, we reach the conclusion that it is only the configuration and momentum observables which go over naturally from classical to quantum mechanics. But, as we remarked earlier, what is perhaps surprising is, not that some classical observables have no quantum analogs, but that any at all do. Nevertheless, one is led to ask why the cut-off occurs between linearity and quadratic-ness in momentum. (Mathematically, it is because different derivative operators only start to become really different after the second derivative.) There is, however, a more direct reason.

Recall that every classical observable generates an infinitesimal canonical transformation. Let us call a canonical transformation "configuration space preserving" if it has the following property: *two points of phase space associated with the same point of configuration space (e.g., (q, p_a) and (q, p'_a)) are taken, by the canonical transformation, to points of phase space which again have this property*. Note that a classical observable is a configuration observable if and only if it assumes the same value on two points of phase space associated with the same point of configuration space. Thus, a canonical transformation is configuration space preserving if and only if it takes configuration observables to configuration observables. The infinitesimal version of this statement is this: *the infinitesimal canonical transformation generated by an observable is to be regarded as configuration space preserving when and only when the Poisson bracket of that observable with any configuration observable is a configuration observable.* But now (37) implies (immediately) that the only observables which generate infinitesimal canonical transformations which are configuration space preserving are the configuration and momentum observables. Thus, these classical observables play a special role.

What is so special about things which are configuration space preserving? Recall that quantum mechanics, as opposed to classical mechanics, requires, not just phase space, but also configuration space itself. Classical observables which "disrupt" configuration space (*i.e.*, whose infinitesimal canonical transformations are not configuration space preserving) might be expected to be awkward in quantum mechanics. Indeed, as we have found, they are awkward; they do not seem to carry over from classical to quantum mechanics.

18. The Hamiltonian. Dynamics

We now wish to describe the passage of our quantum system, with time, from one state to another. In classical mechanics, this description is carried out by introducing a certain observable H, the Hamiltonian. This H then defines the dynamical trajectories, which describe the dynamics. Thus, the most obvious way to proceed is first to try to replace the classical observable H by some sort of analogous quantum observable.

We are confronted immediately by a serious problem. We have seen in Sect. 17 that it is only the configuration and momentum observables which go over directly from classical to quantum mechanics. However, the classical Hamiltonian is almost never a configuration or momentum observable. In fact, it is almost always the case that the Hamiltonian contains quadratic terms in the momenta, $i.e.$, that H takes the form

$$H = g^{ab}p_a p_b + A^a p_a + V \tag{52}$$

Such a classical observable, by Sect. 17, does not go over in the obvious way to a quantum observable.

Fortunately, a remarkably simple result intervenes to save the situation. It is this: *On a manifold M, let g^{ab} be a symmetric (i.e., $g^{ab} = g^{(ab)}$) tensor field which is invertible (i.e., for which there exists a field g_{ab} such that $g^{am}g_{bm} = \delta^a{}_b$). Then, there exists one and only one derivative operator ∇_a on M such that $\nabla_a g^{bc} = 0$.* Proof: Let ∇'_a be any derivative operator on M. We then have, from (6),

$$\nabla_a g^{bc} = \nabla'_a g^{bc} + C^b{}_{am} g^{mc} + C^c{}_{am} g^{bm} \tag{53}$$

We must first show that there is one and only one tensor field $C^a{}_{bc}$ on M such that the right hand side of (53) vanishes. Indeed, there is one, namely,

$$C^a{}_{bc} = -g_{bm}\nabla_c g^{am} - g_{cm}\nabla_b g^{am} + g^{am} g_{bp} g_{cq} \nabla_m g^{pq} \tag{54}$$

as can be checked by substituting (54) into the right hand side of (53). Suppose both $C^a{}_{bc}$ and $\tilde{C}^a{}_{bc}$ caused the right side of (53) to vanish. Then, setting $K_{abc} = g_{am}(C^m{}_{bc} - \tilde{C}^m{}_{bc})$ we have $K_{abc} = K_{a(bc)}$, and, from the vanishing of the right side of (53), $K_{abc} = -K_{bac}$. Using alternatively the first and second of these facts, we have

$$K_{abc} = K_{acb} = -K_{cab} = -K_{cba} = K_{bca} = K_{bac} = -K_{abc}$$

whence $K_{abc} = 0$, whence $C^a{}_{bc} = \tilde{C}^a{}_{bc}$. This completes the proof.

The idea is to use this result to convert the classical Hamiltonian (52) into a quantum observable. In order to do this, we must assume that the g^{ab} in (52) is invertible. (This condition is certainly satisfied for all the usual systems one treats in quantum mechanics.) Thus, if the classical Hamiltonian contains a term quadratic in momentum (together, possibly, with terms linear in momentum and independent of momentum), and if the coefficient of that term, g^{ab}, is invertible, we can introduce the corresponding quantum operator

$$H\psi = (\tfrac{\hbar}{i})^2 g^{ab}\nabla_a\nabla_b\psi + (\tfrac{\hbar}{i})(A^a\nabla_a\psi + \tfrac{1}{2}\psi\nabla_a A^a) + V\psi \qquad (55)$$

where ∇_a in (55) is the unique derivative operator defined by g^{ab} (*i.e.*, such that $\nabla_c g^{ab} = 0$). We needn't be concerned with whether or not the right side of (55) is independent of derivative operator (it isn't), because the classical Hamiltonian prefers one derivative operator. Note that since $\nabla_a g^{bc} = 0$, the first term on the right in (55) is the same as $(\tfrac{\hbar}{i})^2\nabla_a(g^{ab}\nabla_b\psi)$, and the same as $(\tfrac{\hbar}{i})^2\nabla_a\nabla_b(g^{ab}\psi)$. (In more conventional language, with this choice of derivative operator, there is no factor ordering problem.) That is to say, if T^{ab} in (51) is replaced by g^{ab}, the last two terms on the right vanish. It is now immediate that from the discussion following (51) that the operator (55) is self-adjoint.

Thus, if the classical Hamiltonian is of the form (52), with g^{ab} invertible, we can introduce a unique corresponding quantum Hamiltonian (55). As far as I am aware, it is only when the classical Hamiltonian satisfies these conditions that one can "quantize" a classical system. Perhaps one should take the point of view that systems (*i.e.*, quantum theory) operate only within this regime. What is then somewhat mysterious is why it should be that, when one passes to the classical limit, the description has a very natural generalization (*i.e.*, to Hamiltonians other than (52)) which is completely lost within quantum mechanics itself. Alternatively, one could take the following view. In quantum mechanics, there is a Hamiltonian observable, which is of finite differential order. A system described by such a Hamiltonian can, in the classical limit, be described by a classical Hamiltonian algebraic in the momenta. It just happens that it is only when the original quantum Hamiltonian was of second differential order that one can pass uniquely back from classical to the quantum description. But, a priori, one might not have expected to be able to pass from the classical description, uniquely, to the quantum description in any case. It just happens that, in the quadratic case, no essential information gets lost in the classical limit. (Nature, however, tries to confuse us by making most quantum Hamiltonians of second differentiable order, so we get used to the idea that we should be able to recover the quantum description from its classical limit.)

A useful simplification becomes available in the presence of the Hamiltonian (55). We claim: *there are just two alternating tensors (differing only in sign) satisfying:*

$$g_{a_1 b_1}\cdots g_{a_n b_n}\epsilon^{a_1\ldots a_n}\epsilon^{b_1\ldots b_n} = n!$$

That the left side of this equation is nonzero for any nonzero alternating tensor follows from the fact that g_{ab} is invertible. (This is the tensor statement of the fact from matrix algebra that an invertible matrix has nonzero determinant.) But any multiple of an alternating tensor is an alternating tensor: hence, there exists an alternating tensor satisfying the above. That there are just two, differing in sign, follows from the fact that, if ϵ and $\alpha\epsilon$ satisfy the above, then $\alpha^2 = 1$, whence $\alpha = \pm 1$. Thus, our claim is proved. We can use this preferred alternating tensor (pick one) to reduce density fields to tensor fields. Thus, instead of a density ψ (a mapping from alternating tensors to scalars), we can evaluate this ψ on our preferred alternating tensor $\epsilon^{a_1 \cdots a_n}$, to obtain a complex scalar field $\psi(\epsilon)$. Similarly, all tensor densities can be "evaluated on ϵ" to yield tensor fields. Note, furthermore, that, taking the derivative of the equation above, using the derivative operator defined by g^{ab}, we obtain $\nabla_a \epsilon^{a_1 \cdots a_n} = 0$. Hence, $\lambda_a = 0$ in (39). That is to say, if we use the derivative operator defined by g^{ab}, and evaluate tensor densities on the preferred alternating tensor to obtain tensor fields, we can replace derivatives of such densities by derivatives of the corresponding tensor fields. (In other words, the operations *"evaluate on the preferred alternating tensor"* and *"take the derivative by the preferred derivative operator"* commute.)

Our final task is to write down the equation describing the dynamics of our system. At each time t, the system is to be in some state, $\psi(t)$, where $\psi(t)$ is (by the previous paragraph) a complex scalar field on configuration space C, for each t. The dynamical equation, called **Schrodinger's equation**, is

$$- \left(\tfrac{\hbar}{i} \right) \frac{d}{dt} \psi = H\psi \qquad (56)$$

This gives the time rate of change of ψ in terms of ψ at one instant of time. Hence, (56) determines $\psi(t)$ for all t, given $\psi(t_0)$ at one fixed time, t_0.

19. Higher Order Observables–Revisited

We have seen in Sect. 17 that classical observables which are of order quadratic or higher in momentum do not lead to quantum observables. The difficulty was this: *no expression could be found for the quantum observable which was independent of the choice of derivative operator.* But, in Sect. 18, we observed that the presence of a (classical) Hamiltonian quadratic in momentum leads, among other things, to a natural, unique derivative operator. Thus released from what was the primary difficulty in Sect. 17, we return to the question of that section.

We begin with the quadratic case. Let $T^{ab}p_a p_b$ be a classical observable. Then, using the unique derivative operator defined by the Hamiltonian, we have the following corresponding quantum observable

$$\tfrac{1}{2}(\tfrac{\hbar}{i})^2[T^{ab}\nabla_a\nabla_b + (\nabla_a\nabla_b T^{ab}) + \gamma(\nabla_b T^{ab})\nabla_a] \tag{57}$$

where γ is any real number. This operator is self-adjoint for all γ, but, of course, is independent of the choice of ∇_a for no γ. In short, one has, in the presence of a preferred derivative operator, a one-parameter family of candidates for the quantum version of the observable $T^{ab}p_a p_b$. It is clear that this situation continues into higher-order observables. A classical observable of quadratic or higher order in momentum leads, in the presence of a preferred derivative operator, to a family of "corresponding" quantum observables depending on some arbitrary constant parameters.

Thus, one before had no quantum versions of higher-order classical observables —one now has many. It is natural, in such a situation, to attempt to impose some additional conditions on our quantum observables in order to obtain uniqueness. There is a natural such condition: one could require that $(\tfrac{\hbar}{i})$ times Poisson bracket of classical observables correspond to the commutator of the corresponding quantum observables. Consider, for example, the classical observables $T^{ab}p_a p_b$ and $\xi^a p_a$, with Poisson bracket $(2T^{m(a}\nabla_m\xi^{b)} - \xi^m\nabla_m T^{ab})p_a p_b$ (Sect. 12). The corresponding quantum observables are to be (57) and $(\tfrac{\hbar}{i})(\xi^a\nabla_a + \tfrac{1}{2}\nabla_a\xi^a)$. We ask that the commutator of these two observables be $(\tfrac{\hbar}{i})$ times the operator obtained by replacing T^{ab} in (57) by $(2T^{m(a}\nabla_m\xi^{b)} - \xi^m\nabla_m T^{ab})$. The problem is to choose the constant γ in (57) so that this will be the case. The result of this straightforward and tedious calculation is that no choice of γ does the job. *There is, apparently, no way to assign quantum versions to algebraic classical observables, in the presence of a preferred derivative operator, so that commutators correspond to Poisson brackets.*

Having made this observation, we next observe that it is the answer to what is perhaps a somewhat artificial question. This question of the relation between classical and quantum observables can be reformulated as follows. On a manifold C, a **differential operator** consists of $(\nabla_a, T^{a_1 \cdots a_s}, T^{a_1 \cdots a_{s-1}}, \ldots, T^a, T)$, where ∇_a is a derivative operator on C, and the T's are symmetric tensor fields on C. The integer s (if $T^{a_1 \cdots a_s}$ is nonzero) is the order of the differential operator. We can regard differential operators as acting on scalar densities ψ of weight $+\frac{1}{2}$ as follows:

$$(\tfrac{\hbar}{i})^s [T^{a_1 \cdots a_s} \nabla_{a_1} \ldots \nabla_{a_s} + \cdots + T^a \nabla_a + T] \psi \tag{58}$$

Now suppose, in (58), that we rewrite this expression in terms of some other derivative operator ∇'_a, related to ∇_a via $C^a{}_{bc}$ (Eqn. 40). Then (58) becomes

$$(\tfrac{\hbar}{i})^s [T'^{a_1 \cdots a_s} \nabla'_{a_1} \ldots \nabla'_{a_s} + \cdots + T'^a \nabla'_a + T'] \psi \tag{59}$$

where each $T'^{a \cdots c}$ is expressed in terms of the T's, $C^a{}_{bc}$, and ∇_a. But (59) is to be regarded as the action of the differential operator $(\nabla'_a, T'^{a_1 \cdots a_s}, T'^{a_1 \cdots a_{s-1}}, \ldots, T'^a, T')$ on ψ. We regard these two differential operators as equivalent. (More precisely, a differential operator is an equivalence class of objects $(\nabla_a, T^{a_1 \cdots a_s}, T^{a_1 \cdots a_{s-1}}, \ldots, T^a, T)$, where two such objects are equivalent if their actions, (58), on any density ψ of weight $+\frac{1}{2}$, are identical.)

The quantum observables include self-adjoint differential operators. The classical limit of the quantum observable (58) is, of course, $T^{a_1 \cdots a_s} p_{a_1} \cdots p_{a_s}$. Thus, the quantum theory itself has no difficulty in any case with observables. There are many of them lying around, including, presumably, a Hamiltonian observable (possibly quadratic, but also possibly some other order in momentum). We, however, insist on asking about the possibilities for recovering various quantum observables from their classical limit. The situation then becomes quite complicated. Nature, however, doesn't care whether or not quantum observables are recoverable from classical limits.

20. Heisenberg Formulation

What we have been discussing so far is called the Schrodinger formulation of quantum mechanics. There is a second, more natural but essentially equivalent mathematically, formulation called the *Heisenberg formulation of quantum mechanics*. We now introduce it.

It is convenient, in order to illustrate what is involved here, to first return to classical mechanics. Recall that the states in classical mechanics were the points of phase space Γ_C, while the observables were scalar fields on phase space. That is to say, the observables were real-valued functions on the space of states. We may call this the *Schrodinger formulation of classical mechanics*. By contrast, the *Heisenberg formulation of classical mechanics* would be this. The states would be the dynamical trajectories on phase space, *i.e.*, the curves $\gamma(t)$ on phase space, labeled by the time t, whose tangent at each point is the Hamiltonian vector field evaluated at that point. Thus, in the Schrodinger formulation of classical mechanics, the system possesses a state (a point of phase space) at each instant of time: as time marches on, the state changes. In the Heisenberg formulation, a system has one state (a dynamical trajectory), not at each instant of time, but once and for all. Although the system evolves, it always retains the same state, in the Heisenberg formulation. One summarizes the state of affairs by saying that, in the Schrodinger formulation, the state of the system depends on time, while in the Heisenberg formulation, the state is independent of time.

In the Schrodinger formulation of classical mechanics, an observable is a real-valued function on the space of (Schrodinger) states. Similarly, in the Heisenberg formulation, an observable is a real-valued function on the space of (Heisenberg) states. One could, of course, make the space of Heisenberg states into a manifold, so the Heisenberg observables would functions on this manifold. It is convenient, however, to proceed in a slightly different way. Let A be a Schrodinger observable, so A is a scalar field on phase space. We introduce a Heisenberg observable, $O(A, t_0)$, which depends both on A and a choice of time, t_0. This observable is defined as follows: it assigns to the Heisenberg state $\gamma(t)$ (a dynamical trajectory) the number $A(\gamma(t_0))$. That is to say, $O(A, t_0)$ assigns to the Heisenberg state $\gamma(t)$ the number assigned by the Schrodinger observable A to the Schrodinger state $\gamma(t_0)$. Thus, a single Schrodinger observable leads to a one-parameter family $O(A, t)$ (labeled by the parameter t) of Heisenberg observables.

To summarize, a Heisenberg state is a certain one-parameter (labeled by t)

family of Schrodinger states. Whereas the Schrodinger state of a system changes
with time (as the system evolves), the Heisenberg state does not change with
time. In either case, the observables are real-valued functions on the space of
states. Each Schrodinger observable leads to a one-parameter family (labeled by
t) of Heisenberg observables.

Since the Heisenberg observables which arise from Schrodinger observables
depend on the parameter t, it is natural to ask for their rate of change with
respect to t. Intuitively, the difference between $O(A, t_0 + \Delta t_0)$ and $O(A, t_0)$,
acting on a state $\gamma(t)$, is the difference between A evaluated at the point $\gamma(t_0 + \Delta t_0)$ and the point $\gamma(t_0)$ of phase space. But the Hamiltonian vector field is the
tangent vector to a dynamical trajectory. It is immediate, therefore, from (17)
that

$$\frac{d}{dt}O(A, t) = O([H, A], t) \tag{60}$$

This equation describes the time-dependence of Heisenberg observables which
arise from Schrodinger observables. It also shows, for example, that $O(H, t)$ is
independent of t.

Of course, the "reformulation" of classical mechanics above adds practically
nothing to the content if classical mechanics. It is of interest because it provides
an almost perfect analogy for the Heisenberg reformulation of Schrodinger quantum mechanics (which, again, adds practically nothing to quantum mechanics.)

In the Schrodinger formulation of quantum mechanics, a state is a density
ψ on configuration space C, of weight $+\frac{1}{2}$, which is square-integrable. In the
Heisenberg formulation, a state is a solution, $\psi(t)$, of Schrodinger's equation,
(56). That is to say, a Heisenberg state is a certain one-parameter family (labeled
by t) of Schrodinger states. In the Schrodinger formulation, a system possesses
a state at each instant of time, the change in that state with time describing the
evolution of the system with time. In the Heisenberg formulation, the system
possesses just one state once and for all, the "time evolution" of the system
being inherent in the state. Thus, the Schrodinger state describing a system is
time-dependent, while the Heisenberg state is time-independent.

In the Schrodinger formulation, an observable is a self-adjoint operator on
the space of (Schrodinger) states. In the Heisenberg formulation, an observable
is a self-adjoint operator on the space of (Heisenberg) states.

Let A be a Schrodinger observable, so A is a self-adjoint operator on the
space of Schrodinger states. We associate with this A a one-parameter family,
$O(A, t_0)$, (labeled by time t_0) of Heisenberg observables. The definition is this:
if $\psi(t)$ is a Heisenberg state, $O(A, t_0)$, acting on this $\psi(t)$, is that Heisenberg
state $\psi'(t)$ such that $\psi'(t_0) = A\psi(t_0)$. In other words, the action of $O(A, t_0)$
is this. Given a Heisenberg state $\psi(t)$, we "freeze" this state at t_0 to obtain
a Schrodinger state, $\psi(t_0)$. We then act on this Schrodinger state with A to
obtain another Schrodinger state. We then evolve this Schrodinger state to
other times (*i.e.*, other than t_0) using the Schrodinger equation. The result
is another Heisenberg state. Note that we are forced to a definition of this
type. We could not simply take the Heisenberg state $\psi(t)$, and consider $A\psi(t)$
(*i.e.*, the one-parameter family of states obtained by acting on each of the one-

parameter family of states $\psi(t)$ with A), for $A\psi(t)$ would not in general satisfy the Schrodinger equation, *i.e.*, would not in general be a Heisenberg state.

To summarize, a Heisenberg state is a certain one-parameter family (labeled by t) of Schrodinger states. An evolving system is described, in the Schrodinger formulation, by a time-dependent Schrodinger state, and, in the Heisenberg formulation, by a single (time-independent) Heisenberg state. Observables, in either case, are self-adjoint operators on the space of states. Every Schrodinger observable defines a one-parameter family of Heisenberg observables.

Finally, we wish to derive the quantum equation analogous to (60). Fix a Schrodinger observable A, and a Heisenberg state $\psi(t)$ (a one-parameter family of Schrodinger states). For each real number w, $O(A, w)$, acting on the Heisenberg state $\psi(t)$, yields another Heisenberg state $\phi_w(t)$. We can just as well write this as $\phi(w, t)$, a two-parameter family of Schrodinger states. Since $\phi_w(t)$ is a Heisenberg state for each w, we have

$$-\left(\tfrac{\hbar}{i}\right)\frac{\partial}{\partial t}\phi(w, t) = H\phi(w, t) \tag{61}$$

for each w. By definition of $O(A, w)$, we have

$$\phi(t, t) = A\psi(t) \tag{62}$$

for each t. Clearly, (61) and (62) define the two-parameter family of Schrodinger states uniquely in terms of the one-parameter family $\psi(t)$. But, from the diagram below,

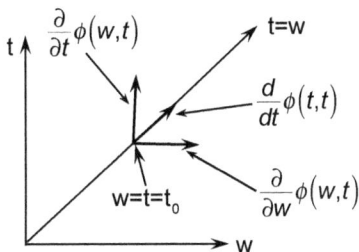

we see that

$$\left.\frac{\partial}{\partial w}\phi(w, t)\right|_{t_0} = \left.\frac{d}{dt}\phi(t, t)\right|_{t_0} - \left.\frac{\partial}{\partial t}\phi(w, t)\right|_{t_0}$$

$$= \left.\frac{d}{dt}(A\psi(t))\right|_{t_0} - \left.\left(-\tfrac{i}{\hbar}H\phi(w, t)\right)\right|_{t_0}$$

$$= -\tfrac{i}{\hbar}AH\psi(t_0) + \tfrac{i}{\hbar}HA\psi(t_0)$$

where, in the last step, we have used (61), (62), and Schrodinger's equation on $\psi(t)$. It is immediate from this equation that

$$\frac{d}{dt}O(A, t) = \tfrac{i}{\hbar}O([H, A], t) \tag{63}$$

This is the desired formula. It gives the rate of change with time of the time-dependent Heisenberg observable arising from any Schrodinger observable A. Note that it implies, in particular, that the Heisenberg observable $O(H,t)$ is independent of t.

It is of interest to work out the right side of (61) for our (Schrodinger) configuration and momentum observables. We take, for our Hamiltonian, (55). Let α be a scalar field on configuration space, and $Q(\alpha)$ the corresponding (Schrodinger) observable. Then

$$
\begin{aligned}
&(HQ(\alpha) - Q(\alpha)H)\psi \\
&= [(\tfrac{\hbar}{i})^2 g^{ab}\nabla_a\nabla_b + (\tfrac{\hbar}{i})(A^a\nabla_a + \tfrac{1}{2}\nabla_a A^a) + V](\alpha\psi) \\
&\quad -\alpha[(\tfrac{\hbar}{i})^2 g^{ab}\nabla_a\nabla_b + (\tfrac{\hbar}{i})(A^a\nabla_a + \tfrac{1}{2}\nabla_a A^a) + V]\psi \\
&= (\tfrac{\hbar}{i})^2[\psi g^{ab}\nabla_a\nabla_b\alpha + 2g^{ab}\nabla_a\alpha\nabla_b\psi] + (\tfrac{\hbar}{i})[\psi A^a\nabla_a\alpha] \\
&= (\tfrac{\hbar}{i})[P(2g^{ab}\nabla_b\alpha) + Q(A^a\nabla_a\alpha)]\psi
\end{aligned}
$$

Note that this reduces, for the case of a particle in Euclidean space, to the usual Heisenberg equation of motion. The corresponding equation for the momentum operator, $P(\xi)$, is more complicated:

$$
\begin{aligned}
&(HP(\xi) - P(\xi)H)\psi \\
&= [(\tfrac{\hbar}{i})^2 g^{ab}\nabla_a\nabla_b + (\tfrac{\hbar}{i})(A^a\nabla_a + \tfrac{1}{2}\nabla_a A^a) + V][(\tfrac{\hbar}{i})(\xi^c\nabla_c + \tfrac{1}{2}\nabla_c\xi^c)]\psi \\
&\quad - [(\tfrac{\hbar}{i})(\xi^c\nabla_c + \tfrac{1}{2}\nabla_c\xi^c)][(\tfrac{\hbar}{i})^2 g^{ab}\nabla_a\nabla_b + (\tfrac{\hbar}{i})(A^a\nabla_a + \tfrac{1}{2}\nabla_a A^a) + V]\psi \\
&= 2(\tfrac{\hbar}{i})^3(g^{m(a}\nabla_m\xi^{b)})\nabla_a\nabla_b\psi + \ldots
\end{aligned}
$$

where the dots denote a rather complicates first-order differential operator. This Heisenberg equation of motion is normally only written when ξ^a is a symmetry of the geometry of configuration space, i.e., when $\pounds_\xi g^{ab} = \xi^m\nabla_m g^{ab} - g^{mb}\nabla_m\xi^a - g^{am}\nabla_m\xi^b = -2g^{m(a}\nabla_m\xi^{b)} = 0$. In this case, the highest order term on the right above vanishes, and the formula reduces to

$$
(HP(\xi) - P(\xi)H)\psi = -(\tfrac{\hbar}{i})[P(\pounds_\xi A^a) + Q(\pounds_\xi V)]\psi
$$

which, of course, reduces to the familiar formula in standard examples (e.g., particle in Euclidean space). These calculations are merely intended to illustrate, for special cases, the general formula (63).

21. The Role of Observables (Classical Mechanics)

Our approach heretofore has been to introduce the states before the observables. Thus, in classical mechanics, observables were functions on the space of states; in quantum mechanics, operators on the space of states—in both cases, observables were defined in terms of their relationship with states. It is natural to proceed in this way, for one thinks of a system as possessing a state (independently of any observations which might be made on it), with the observations representing things which are applied to these states to give us information about the state. One could, however, imagine an alternative point of view. One could regard states, not as attributes of a system somehow handed down from above, but rather as something which must be determined by means of observations. In other words, one could take a more observer-oriented point of view, in which the things which involve us observers directly—the observables—play the fundamental role. This point of view would be reflected within the mathematical formulation by a formalism in which the observables are introduced as the fundamental objects, and the states only as subsidiary objects defined in terms of the observables. As an example, we first carry out such a reformulation for classical mechanics.

We suppose that we are given some system to study. We have some mechanism (*e.g.*, hitting with a stick) by which we can cause our system to change state, so we can experiment with various states. We are to be furnished, furthermore, with an enormous basket full of observables, where we think of an observable as a box from which there protrudes a probe (which can be brought in contact with the system), and which has a meter (which reads some value when the system is in interaction with the observable). Our job is to introduce a description of what is going on in terms of the meter readings of the observables.

One first, by trial and error, makes the following basic observation: if, to the system, observable A, then B, and then A are applied in quick succession (so the system does not evolve appreciably between observables), then the readings obtained for the two A observations coincide. We interpret this to mean that the application of our observables does not disturb the state of the system (*i.e.*, the application of B leaves invariant the A-reading, for all As, and hence leaves the state alone). It follows that the application of several observables in quick succession is independent of the order in which the observables are applied.

81

Two observables, A and A', will be said to be *equivalent* if they always give the same result, applied to our system (more precisely, if the application of A and A' in quick succession yields the same reading on the two meters). We write equivalence of observables with an equals sign.

We lose nothing in generality, yet simplify the discussion, by assuming that our basket of observables has three properties. A *constant observable* is one whose meter reading is always the same number (no matter what the system is doing). We assume that, for each real number, there is in our collection of observables a constant observable which always yields that number. (If some constant observable were missing from our collection, it would be easy to construct one to add to the collection. Just take an empty box, and paint on it a dial and needle always reading, for example, "3".) Now let A and B be two observables. We define a new observable, obtained by applying A and B to the system in quick succession, and adding the readings on the two meters. This new observable, called the *sum of observables* A and B, will be written $A + B$. Similarly, the *product of two observables*, AB, is another observable. We now assume that our collection of observables contains, along with any two observables, both their sum and their product. (If some sum, *e.g.*, $A+B$, were missing from the collection, we could always construct, from A and B, $A + B$ and add it to our collection). Note that addition and multiplication of observables satisfy the same commutative, associative, and distributive laws that addition and multiplication of real numbers do.

An **algebra** $\mathscr{A}$ is a vector space on which one or more bilinear products is defined. (Thus, let $\mathscr{A}$ be a vector space on which products AB, $A \times B$, and $[A, B]$ are defined, satisfying $(A+cB)C = AC+cBC$, $A(B+cC) = AB+cAC$, for c a number, and A, B, C in $\mathscr{A}$, and similarly for the other two products. Then $\mathscr{A}$ is an algebra.)

The discussion above can now be summarized by the statement that we might as well assume that our collection $\mathscr{A}$ of observables forms an algebra (with just one product, which is also associative and commutative).

A *state* is to be characterized in terms of what the observables have to say about it. That is to say, a **state** is a mapping $\sigma : \mathscr{A} \to \mathbb{R}$, from our algebra of observables to the reals, satisfying the following conditions:

1. If C is the constant observable with value c, then $\sigma(C) = c$.

2. If A and B are observables, then $\sigma(A + B) = \sigma(A) + \sigma(B)$.

3. If A and B are observables, then $\sigma(AB) = \sigma(A)\sigma(B)$.

These three conditions, of course, reflect the operational meaning of *constant observables, addition of observables, and multiplication of observables.*

This is nearly the end of the story. From the algebra of observables, one introduces the states. We have still to ask, however, whether or not we obtain, in this way, the "correct" collection of states. The answer depends on whether or not we have enough observables in our original collection $\mathscr{A}$. For example, if $\mathscr{A}$ consisted of just the constant observables, then we would obtain, by the construction above, just one state. Suppose, however, that all observables were included in our collection. Since we "really know" that our system has a phase space, and the observables are scalar fields on phase space, this amounts to the assumption that for every scalar field on phase space there is an instrument which measures that scalar field. We can then ask: *does our construction above yield the correct collection of states?* First note that every point of phase space does indeed define a mapping from observables to the reals (namely, evaluation of the scalar field at the point). Hence, every point of phase space does define a state according to our construction. What needs to be shown is that no additional states—no points of phase space—have been introduced. This is a consequence of the following:

Theorem. Let M be a manifold, and let $\mathscr{A}$ be the algebra of scalar fields on M. Let $\sigma : \mathscr{A} \to \mathbb{R}$ satisfy the three conditions above for a state. Then there exists a point p of M such that $\sigma(A) = A(p)$ for every A in $\mathscr{A}$.

"proof": the only proof I know requires the Whitney embedding theorem, a standard but rather technical result in differential geometry. Since a proof in the special case $M = \mathbb{R}^n$ illustrates the essential idea, and since a proof in this special case, together with the embedding theorem, yields easily a proof in general, we shall assume $M = \mathbb{R}^n$.

Let σ be a mapping as above, and consider the scalar fields $x^1, \ldots, x^n$ on M. Set $\underline{x}^1 = \sigma(x^1), \ldots, \underline{x}^n = \sigma(x^n)$. Let p the point with coordinates $(\underline{x}^1, \ldots, \underline{x}^n)$. Then, for f any scalar field on M, we can write $f = \underline{f} + f_1(x^1 - \underline{x}^1) + \cdots + f_n(x^n - \underline{x}^n)$, where $\underline{f}$ is a constant scalar field, and $f^1, \ldots, f^n$ are scalar fields. Then $\sigma(f) = \sigma(\underline{f}) + \sigma(f_1)(\sigma(x^1) - \sigma(\underline{x}^1)) + \cdots + \sigma(f_n)(\sigma(x^n) - \sigma(\underline{x}^n)) = \sigma(\underline{f})$, where, in the last step, we have used that $\sigma(x^i) = \underline{x}^i = \sigma(\underline{x}^i)$. Thus, $\sigma(f) = \sigma(\underline{f}) = \underline{f} = f(p)$, completing the proof.

Thus, if we have all the observables in our original collection, we can construct the space of states as above, and indeed obtain precisely the correct space of states. A similar conclusion holds even if original algebra includes fewer observables. For example, it easy to check, using the same methods above, that, if $\mathscr{A}$ is the algebra of algebraic observables, then we obtain the correct space of states. Note, furthermore, that, if $\mathscr{A}$ includes all the configuration and momentum observables, then, since $\mathscr{A}$ is an algebra, $\mathscr{A}$ necessarily includes all algebraic observables. Then, so long as $\mathscr{A}$ includes all configuration and momentum observables, precisely the correct space of states emerges from the construction above.

22. The Role of Observables (Quantum Mechanics)

Ideally, one would now like to repeat, for quantum mechanics, what we did in the previous section for classical mechanics. That is to say, one would like to introduce a collection of observables (measuring instruments) on a quantum system, subject to some physically reasonably conditions regarding the action of the instruments on the system, and then define the states in terms of the observables. In this way, one might expect both to obtain a more operational setting for quantum mechanics, and to provide more direct physical interpretations for certain objects which appear in the mathematical formalism of quantum mechanics. It appears, however, that certain difficulties intervene to prevent a simple and direct treatment of quantum mechanics along these lines. In this section, we discuss these difficulties.

We must first consider the mechanism by which observables act on states. Within the formalism of quantum mechanics, an observable is a self-adjoint operator on the space of states. What we must do is relate the mathematical action of this operator to the physical action of the observing instrument on the state. There are, apparently, two possible such relationships.

1. *Active observables.* One wishes to apply to the state the observable associated with the self-adjoint operator A. The meter on the instrument reads an eigenvalue of A, *i.e.*, a number such that $A\psi = a\psi$ for some nonzero state ψ. During the observation, the system suddenly changes its state so that, after an observation yielding the value a, the system is in the corresponding eigenstate (*i.e.*, the ψ above). Even though the initial state ψ, as well as the self-adjoint operator A, are known, the final state into which the system passes is not uniquely determined. Rather, what is determined is the probabilities for the various eigenvalue-eigenstate pairs.

2. *Passive observables.* The application, to the state ψ, of the observable associated with the self-adjoint operator A yields (as the reading of the instrument) the expectation value of A in our state, *i.e.*, the real number $\langle \psi \,|\, A\psi \rangle$. The effect of this measurement on the state of the system can be made arbitrarily small.

I believe that it is true to say that, given A, one can construct measuring instruments, of either the active or the passive type, corresponding to A.

The measuring instruments we introduced in classical mechanics were passive, although one could, of course, have introduced also active instruments (which would change the state of the system). Of course, one could, if it should turn out to be convenient, simply require that the instrument he uses be of one or the other type. In particular, one often treats quantum mechanics using only active instruments. The possibilities for defining states in terms of observables depend on which type of observing instruments one chooses to use. We shall consider the results of using both types.

The disadvantage of using active instruments is that they interfere, in a significant and uncontrollable way, with the state being observed. In fact, the original state is for all practical purposes destroyed by the making of a single active observation. However, a vestige of the original does survive in that the original state determines the probability distribution for the various possible eigenvalues which could emerge from the observation. Unfortunately, a single observation on a single system yields, not a probability distribution, but a single eigenvalue. It is clearly impossible to describe a state in terms of observations on that state in the presence of this strong interference of the observing instrument with state itself.

There is, however, a method for avoiding at least certain of these difficulties. Let us imagine that we have been given a method for preparing the system so as to be in a given state. We then make up an infinite collection of copies of our original of copies of our original system, all in the same state. This collection is called an **ensemble**. One now imagines applying observing instruments, not to a single system (in the single state), but to an ensemble. This is done by making the observation on some infinite sub-collection of the systems forming the ensemble, leaving, at the same time, an infinite collection of systems in the ensemble unobserved. (For example, one could observe every other system in the ensemble.) For each system observed, one records the number (eigenvalue) determined by the observing instrument. The observed systems can then be discarded (their states having been ruined). There remains, however, an infinite collection of systems which have not been disturbed, *i.e.*, an ensemble for further study. Thus, by introducing ensembles rather than states, one avoids the difficulties associated with the interference of observations with the state of the system. One must pay for this device, however, in that one must, in a sense, already know what the states are, for, to prepare an ensemble, one must have in hand a repeatable prescription for putting the system in a given state.

In any case, the information we obtain by applying an observable to an ensemble is a **probability measure** on the real line (*i.e.*, a positive measure such that the total measure of the line is one).

Thus, one would attempt to proceed as follows. One would introduce an algebra of (active) observables. States would be defined as mappings from this algebra to probability measures on the real line, such that the mapping satisfy certain conditions. One would then have to show that the correct space of states is obtained in this way. Unfortunately, it appears to be very difficult to make such a program work, in the sense that one finds suitable structure on the observables and suitable additional conditions for the definition of a state, such that the structure and conditions can be motivated physically, and such that the resulting collection of states agrees with the space of states of quantum mechanics.

It is perhaps not too surprising that the use of active observables should lead to difficulties. The program of defining states in terms of observables works extremely well in classical mechanics, and in that case the observables are in every sense passive. One would naturally expect that the passive observables would be the most natural to use also in quantum mechanics.

Let us attempt, then, to describe states in terms of expectation values for self-adjoint operators, *i.e.*, in terms of passive observables. It is perhaps not unreasonable to ask that included in our collection of observables be all configuration and momentum observables. We begin with a mathematical question: *does knowledge of the numbers* $\langle \psi \,|\, Q(\alpha)\psi \rangle$ *and* $\langle \psi \,|\, P(\xi)\psi \rangle$ *for all* α *and* ξ^a *determine* ψ? If not, our goal is hopeless; if so, we can ask how this determination comes about. Set $\psi = \phi e^{i\lambda}$, where ϕ and λ are real scalar fields. (For simplicity, we take ψ as a scalar field.) Then

$$\langle \psi \,|\, Q(\alpha)\psi \rangle = \int \overline{(\phi e^{i\lambda})}\alpha\phi e^{i\lambda} = \int \phi^2\alpha \tag{64}$$

$$
\begin{aligned}
\langle \psi \,|\, P(\xi)\psi \rangle &= \int \overline{(\phi e^{i\lambda})}(\tfrac{\hbar}{i})(\xi^a\nabla_a + \tfrac{1}{2}\nabla_a\xi^a)\phi e^{i\lambda} \\
&= \int (\tfrac{\hbar}{i})\phi e^{-i\lambda}(e^{i\lambda}\xi^a\nabla_a\phi + i\phi e^{i\lambda}\xi^a\nabla_a\lambda + \tfrac{1}{2}\phi e^{i\lambda}\nabla_a\xi^a) \\
&= \int (\tfrac{\hbar}{i})\phi\xi^a\nabla_a\phi + \int \hbar\phi^2\xi^a\nabla_a\lambda + \int \tfrac{1}{2}(\tfrac{\hbar}{i})\phi^2\nabla_a\xi^a \\
&= \int \tfrac{1}{2}(\tfrac{\hbar}{i})(2\phi\xi^a\nabla_a\phi + \phi^2\nabla_a\xi^a) + \int \hbar\phi^2\xi^a\nabla_a\lambda \\
&= \int \tfrac{1}{2}(\tfrac{\hbar}{i})\nabla_a(\phi^2\xi^a) + \int \hbar\phi^2\xi^a\nabla_a\lambda \\
&= \int \hbar\phi^2\xi^a\nabla_a\lambda \\
&= \hbar\langle \psi \,|\, Q(\xi^a\nabla_a\lambda)\psi \rangle \tag{65}
\end{aligned}
$$

Knowledge of the expectation value of $Q(\alpha)$ for all α determines, by (64), ϕ. Then the expectation value of $P(\xi)$ for all ξ^a determines, by (65), $\nabla_a\lambda$, and hence λ up to a constant. Thus, $\psi = \phi e^{i\lambda}$ is determined up to an overall phase (which is all the wave function is ever determined up to, anyway). Our program should, in principle, succeed.

Denote by $\mathscr{Q}$ the collection of all (passive) configuration observables, and by $\mathscr{P}$ the collection of all (passive) momentum observables. One knows, operationally, how to multiply an element of $\mathscr{Q}$ or an element of $\mathscr{P}$ by a real number, and how to add two elements of $\mathscr{Q}$ or two elements of $\mathscr{P}$. Thus, each of $\mathscr{Q}$, $\mathscr{P}$ has the structure of a vector space. A *state* is to consist of linear mappings $\sigma : \mathscr{Q} \to \mathbb{R}$ and $\sigma : \mathscr{P} \to \mathbb{R}$. We wish to impose additional conditions on these mappings in order that $\sigma(\alpha) = \langle \psi | Q(\alpha)\psi \rangle$ and $\sigma(\xi) = \langle \psi | P(\xi)\psi \rangle$ for some state ψ. It follows from the linearity of σ that there exist scalar and vector fields μ and μ_a on configuration space such that

$$\sigma(\alpha) = \int \alpha\mu \tag{66}$$

$$\sigma(\xi) = \int \xi^a \mu_a \tag{67}$$

for all α and ξ^a. (We should, more precisely, allow μ and μ_a to be *distributions* on configuration space. Thus, for example, a scalar distribution is normally defined as a linear mapping from scalar fields on a manifold to reals. In order to avoid technical complications, however, we shall restrict consideration to the case when μ and μ_a are ordinary tensor fields.) We wish to define ϕ by $\phi^2 = \mu$, and for this we need $\mu \geq 0$. We can state this condition in terms of our operators as follows: if $\alpha \geq 0$, then $\sigma(\alpha) \geq 0$. We wish to define λ by $\hbar\phi^2\nabla_a\lambda = \mu_a$, which requires that $\phi^{-2}\mu_a$ be a gradient. This condition is equivalent to the requirement that, whenever ξ^a is such that $\nabla_a(\phi^2\xi^a) = 0$, $\sigma(\xi) = 0$, *i.e.*, to the requirement that whenever ξ^a is such that $\sigma(\xi^a\nabla_a\alpha) = 0$ for all α, $\sigma(\xi) = 0$. With these two conditions, we can recover ϕ and λ, and hence the state $\psi = \phi e^{i\lambda}$, from the mappings σ.

To summarize, we introduce two vector spaces, $\mathscr{Q}$ and $\mathscr{P}$, of observables. A state is then defined as a pair of linear mappings, $\sigma : \mathscr{Q} \to \mathbb{R}$ and $\sigma : \mathscr{P} \to \mathbb{R}$, satisfying two conditions:

1. If $\alpha \geq 0$, then $\sigma(\alpha) \geq 0$.

2. If $\sigma(\xi^a\nabla_a\alpha) = 0$ for all α, then $\sigma(\xi) = 0$.

Then, roughly speaking, for every such state there is a unique wave function ψ on configuration space such that $\sigma(\alpha) = \int \bar{\psi}\alpha\psi$ and $\sigma(\xi) = \int \bar{\psi}(\frac{\hbar}{i})(\xi^a\nabla_a + \frac{1}{2}\nabla_a\xi^a)\psi$.

Note that our two conditions in the definition of a state are indeed satisfied in the classical case. (Classically, the second condition amounts essentially to the statement that vector fields are defined as derivations on scalar fields, so the zero derivation defines the zero vector field). Why were we able to avoid explicit mention of such conditions in the classical case? The reason is that we were able there to introduce a far stronger condition in the definition of states, namely, the product condition: $\sigma(AB) = \sigma(A)\sigma(B)$. One might imagine, therefore, that matters could be simplified in the quantum case by the introduction of a similar product condition. Of course, it is easy, operationally, to introduce the product of two (passive, quantum) observables. Given instruments which measure A and B, one constructs an instrument which measures first A and then B, and multiplies. (Since passive observables do not affect the states, this construction is independent of order.) The difficulty, however, is this: if A and B are self-adjoint operators, then, in general,

$$\langle \psi | A\psi \rangle \langle \psi | B\psi \rangle \neq \psi AB\psi \tag{68}$$

The left hand side is the number associated with the "product observable" constructed above. The right side is the expectation value associated with the product of operators. In short, the operational procedure of taking products of observables does not correspond to the mathematical procedure of taking products of operators. The incorporation of operational products into our quantum observables thus adds essentially nothing new.

Thus, we indeed obtain a formulation of quantum mechanics in which states are described in terms of (passive) observables. Although the two conditions we require in the definition of a state (that $\sigma(\alpha) \geq 0$ if $\alpha \geq 0$, and that $\sigma(\xi) = 0$ if $\sigma(\xi^a \nabla_a \alpha) = 0$ for all α) are perhaps not terribly unnatural physically, the second certainly seems rather artificial. One could easily invent other conditions of the same general type which are not imposed. There is, however, another—perhaps even more unpleasant—feature of this formulation. In the classical case, we could think of our observables as measuring instruments, instruments which we knew how to apply to our system, but about which no further information was required. This is not at all the case for our quantum observables. Firstly, we need to know which instruments measure configuration and which momentum observables. That is to say, each instrument must have, inscribed on its side, either "$\mathscr{Q}$" or "$\mathscr{P}$". In fact, we need, for the construction of states above, much more than this. We also need to know, for each instrument which measures a configuration observable, which scalar field α on configuration space this instrument measures (the expectation value of), and, for each instrument which measures a momentum observable, which vector field ξ^a is observed. This additional information is necessary in order that the construction of states can be carried out. On the other hand, no operational procedure is given for obtaining this additional information, *e.g.*, no procedure for discovering what an instrument measures by taking it apart. In fact, the introduction of such observables requires already a knowledge of what configuration space is—but configuration space determines already the quantum states.

To summarize, because the measurement process in quantum mechanics is apparently more subtle than that in classical mechanics, the simple formulation of classical mechanics in terms of observables apparently does not carry over in a natural and equally simple way into quantum mechanics.

23. The Interpretations of Quantum Mechanics

It is conventional to attempt to formulate what are called *interpretations of quantum mechanics*. It is natural to ask, firstly, why such an interpretation is needed. Apparently, it is because the objects which appear in the mathematical formulation of quantum mechanics (wave functions, self-adjoint operators, *etc.*) are related in a rather tenuous way to the things (*e.g.*, dial readings) which human observers actually see. Consider an example. An atom in some excited state is to decay, emitting a photon. The photon is to ionize an atom in a geiger counter, so the resulting electron will, because of an applied electric field, set up a cascade of electrons in the counter. The resulting electric signal will, after being sent through an amplifier, activate the needle of a meter. We can begin this system with the original atom in an excited state, and ask how the system will evolve. This is a question with which we can, at least in principle, deal, within the context of quantum mechanics. We are to introduce the configuration space of this entire system (consisting of the atom, the geiger counter, the amplifier, and the meter). On this configuration space, we are to introduce an appropriate differential operator, the Hamiltonian. We are then to introduce a certain wave function ψ describing the initial state of the system. Schrodinger's equation will then determine ψ at later times.

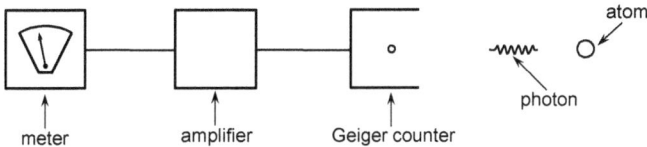

Clearly, the program above is so formidable that one is not likely to carry it out in practice. One can, however, gain insight into what a solution would look like by considering a simpler problem having roughly similar features. Consider a system with 2-dimensional configuration space, the xy-plane. Let the classical Hamiltonian be

$$H = p_x{}^2 + x^2 + p_y{}^2 + y^2 + f(x - y)$$

where f is some function of one variable. Thus, when f is zero, we have two independent one-dimensional harmonic oscillators, f represents a coupling between two such oscillators. We write down the quantum description of this system, and begin in the state $\psi(x, y) = \psi_g(x)\psi_e(y)$, where $\psi_g(x)$ is the ground state, and $\psi_e(y)$ some excited state, of a one-dimensional harmonic oscillator. We now allow this state to evolve, by Schrodinger's equation, and see what happens. What happens is exactly what one expects to happen: the energy "leaks" slowly, because of the coupling, from one oscillator to the other. That is to say, after a short time, the first oscillator is nearly in its ground state, but with a small admixture of excited states. As time goes on, the amount of this admixture slowly increases.

We can think of this simple system as analogous to the more complicated system which preceded it. Initially, the system was described by a wave function such that the atom was in an excited state, and the needle on the meter reads zero. As time goes on, the component of the wave function of the atom corresponding to the ground state increases, while the component of the wave function of the needle corresponding to a nonzero meter reading also increases. That is to say, one would expect that the state of our system will evolve, slowly and continuously, from one in which the atom is excited and the meter reads zero, to one in which the atom has decayed and the meter reads nonzero.

On the other hand, one can actually buy, *e.g.*, from Sears, a geiger counter, amplifier, meter, and excited atom. One sets up the experiment, and watches the needle as the system evolves. What one actually sees, of course, is that the needle remains at zero for a while, then jumps to some value. One often says that, at the instant the needle jumped, the atom decayed, and the photon was received by the geiger counter, amplified, and recorded by the meter.

We now have two descriptions of this experiment. In one, via the formalism of quantum mechanics, the wave function of the needle continuously evolves from that corresponding to a zero reading to that corresponding to a finite reading. In the other, via direct observation, the needle is seen, not in some "smeared out state", half reading zero and half something else, but is rather seen to remain at zero for a while then jump suddenly. The problem of interpretation, as it bears on this example, is to somehow bring together these two descriptions. That is to say, the problem of interpretation is to make more sharp the relationship between the objects which appear in the formalism of quantum mechanics and the things observers actually see.

It is not at all uncommon in physics (perhaps it is even characteristic!) that the objects at the center of the mathematical description are not the first and most obvious things observers see. Special relativity is a good example. Fundamental in the mathematical description of special relativity are the events (instantaneous occurrences at a point), which are assembled into the four-dimensional space-time manifold. Yet, although each of us lives (ignoring gravity) in the four-dimensional space-time manifold of special relativity, we do not experience any four-dimensional manifold stretching out before us. Instead, we experience three-dimensional space, and the passage of time. Why, then, does one not feel a need to acquire an "interpretation" of special relativity?

It is my impression that an interpretation is as necessary in special relativity as in quantum mechanics, but that, in the case of the former, an interpretation is available and widely accepted, so that its role as an interpretation normally goes unemphasized. This interpretation is that each of us (our past, present, and future) is represented by a world-line in space-time, that what we experience is light rays reaching our world-line from distant events, objects whose world-lines meet ours, *etc.* Within this framework—a framework in which the observer is incorporated directly—one can, in some sense, account for what we as observers actually see.

It is perhaps not the whole truth to say that those experiences for which special relativity might give account are in fact accounted for by special relativity. It is, for example, not so easy to account, within special relativity, for the fact that one cannot go back and actively experience his past. However, one perhaps takes the point of view that these finer details are somehow to be described in terms of the detailed physics of what goes on in the brain. One might also say, based on this discussion of special relativity, that the purpose of an interpretation is, not really to explain anything, but rather to put one at ease while he works within the mathematical formalism.

We return to quantum mechanics. As discussed above, there is a sense in which quantum mechanics is inadequate. On the other hand, there is certainly a sense in which quantum mechanics is far more than adequate: *it makes beautiful and accurate predictions in remarkable agreement with Nature.* One would therefore not like to change in a substantive way the structure of quantum mechanics, *e.g.*, change its mathematical formalism. What one would like to do is to look at quantum mechanics from a different point of view, or at least utter some reassuring words about it, so that the apparent inadequacy seems less serious. That is to say, what one would like to do is formulate an interpretation of quantum mechanics.

24. The Copenhagen Interpretation

The most widely accepted interpretation of quantum mechanics is what is called the **Copenhagen interpretation**.

The idea of the Copenhagen interpretation, as I understand it, is that quantum mechanics is fine as far as it goes, but that it does not go far enough. What is missing is the relationship between the wave function and what observers see. The Copenhagen interpretation attempts to provide this additional link.

One first notices that, if you ask an observer what it was that he actually saw, he will say "I saw the meter read 3.5.", or "I saw the light go on.", but never "The wave function of the needle on the meter was the following superposition...". Thus, this additional link is to somehow tie together quantum phenomena (*i.e.*, wave functions) and classical phenomena (*i.e.*, points of configuration space).

Consider again the experiment on page 91. We wish to regard the meter classically, and the original atom quantum mechanically. We therefore introduce a break at some point between the chain of instruments linking the atom to the meter, *e.g.*, between the atom and the geiger counter. Everything up to the break (in the case, just the atom itself) will be treated quantum mechanically, and everything beyond the break (in this case, the geiger counter, amplifier, and meter) will be treated classically. Put more formally, we introduce a classical phase space to describe the states of the meter-amplifier-geiger counter system, and a space of quantum states (*i.e.*, complex functions on an appropriate configuration space) to describe the original atom.

The next step is to introduce a coupling of these two systems, across the break. This is done by assigning a self-adjoint operator A to the quantum system, and, at the same time, assigning a state of the classical system to each eigenvalue of A. The prescription for making these assignments is not specified by the interpretation, but the assignments are to be so made that, in the end, one obtains an appropriate description of this measurement process. The essential assumption is that one can introduce a break in this chain of instruments, and describe the information flow across the break in terms of a self-adjoint operator A on the quantum system, and states of the classical system labeled by eigenvalues of A.

The break comes into play during what is called *process of measurement*. The following then takes place. The Schrodinger equation for the quantum system, and the Hamilton equation for the classical system are, for a moment, suspended. During this suspension the quantum system is thrown into an eigenstate of A, and, simultaneously, the classical system is thrown into the state associated with the corresponding eigenvalue of A. The process of measurement is thus a process of sudden change in state. In this way, the formalism of quantum mechanics is to be brought into contact with the (classical) things people, as observers, see.

The particular eigenstate of A into which the quantum state is thrown during the process of measurement is not specified precisely. Rather, what can be determined is the probability distribution for various of the probabilities. Thus, there is, in a sense, a loss of determinism in the Copenhagen interpretation, associated with the process of measurement. It is essentially at this point that probability is brought into quantum mechanics.

One might object to the interpretation above on, among others, the following grounds. No prescription has been given for where to introduce the break between the quantum and classical systems. Thus, in our example, one could as well have drawn the break between the geiger counter and amplifier, or between the amplifier and meter. It is, so I understand, part of the lore of the Copenhagen interpretation that the result (i.e., the resulting probability distribution for the meter readings) is independent of this choice. It is not completely clear how one formulates a precise statement along these lines, although indications are (e.g., from simple examples) that something of this general type is true. Thus, one is apparently free to introduce the break at any point of the chain with the same result, i.e., with the same probability distribution for (classical) meter readings.

One might now object on somewhat different grounds. Since one is free to introduce the break wherever one wishes, it is prescribed, within the interpretation, whether or not a given link in the chain of measurement instruments is quantum. Thus, for example, the geiger counter could be regarded either as a classical or as a quantum system, depending on where the (arbitrary) break was introduced. The Copenhagen answer would be this. If you are concerned about the geiger counter, make a measurement on it. Thus, one would introduce a tributary in the chain of measurements, with additional instruments looking in particular at the geiger counter.

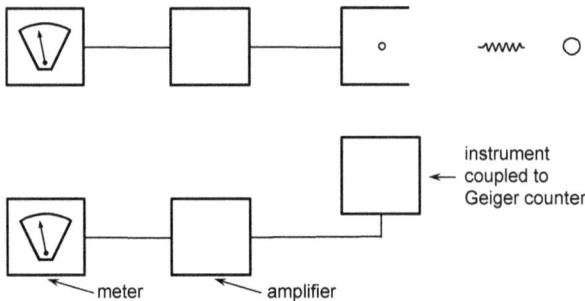

Now one has a new system of linked instruments. One is supposed to apply to it the Copenhagen prescription, i.e., one is to introduce a break in the chain, with the two sides linked by a self-adjoint operator, *etc.* The result will be a probability distribution for classical results (at the end of the chain), a distribution independent of where the break was drawn. In any case, one will obtain an unambiguous result. Whether one chooses to interpret this result as some statement about whether the geiger counter is quantum or classical is certainly one's option. But, in any case, no difficulty arises from the failure of the Copenhagen interpretation, in the first experiment, to specify whether the geiger counter was "classical" or "quantum".

The claim of the interpretation is just that

if you specify to it with sufficient precision what you will do in making a measurement, it will tell you the probability distribution for the results of the measurement.

It does precisely this in every case. It does not have to deal with what "really happens". It is, in some sense, "consistent".

One could, equally well, have expressed this interpretation in a somewhat different way. One could as well have asserted, at some point, that the entire system is in fact subject to the laws of quantum mechanics. That is to say, one could have asserted that, in fact, there is no break in the system. Then the Copenhagen prescription could have been regarded merely as the rules for carrying out the calculation giving the probability distribution for the results of the measurement. That is to say, one could have presented matters so that the break business was presented as a computational tool rather than a statement about what "really happens". Clearly, nothing of any real substance (*e.g.*, numerical conclusions) depend on this mode of formulation. This "modified Copenhagen interpretation" is, however, perhaps more natural sounding, for it requires only that we in our calculations do strange things (*e.g.*, introduce a break), rather than that Nature do them.

25. The Everett Interpretation

There is an alternative interpretation of quantum mechanics, called the **Everett interpretation**, which we now describe.

It is convenient to begin with some remarks about classical mechanics. The formalism of quantum mechanics was introduced originally, in order to describe rather implicitly defined things called *systems*. Let us, just to see what will result, attempt to push this formalism into a context broader than that for which it was originally intended. What we wish to do is to treat the entire Universe as a single system within the formulation of classical mechanics. Thus, we suppose that there is some configuration space C for our Universe. Let Γ_C be the corresponding cotangent bundle. We suppose that there is some scalar field H on this Γ_C, the Hamiltonian. In the beginning (*i.e.*, at time $t = 0$), the Universe occupied some point of phase space. The future evolution of the Universe is then determined by Hamilton's equation, *i.e.*, by the condition that the dynamical trajectory describing the evolution of the Universe be an integral curve of the Hamiltonian vector field. We emphasize that within this description is to be included the *entire* Universe—you and I, all observers, all observing instruments, everything.

One might feel uncomfortable about the above for the following reason. In our original formulation of classical mechanics, one imagined an external observer who, by manipulation and examination of the system under study, was able to assign it a configuration space and Hamiltonian. But the present context excludes such external observers. Thus, for example, consider a point of phase space through which the dynamical trajectory of our universe does not pass. What does it mean to speak of this "possible state of the Universe" when the Universe never, in fact, occupies that state?

It seems to me that the extrapolation involved here is, essentially, no better and no worse than the extrapolations one always makes in physics. Suppose, for example, that an external observer wishes to study a classical system. He has some means to manipulate the system, in order to study its states. Nonetheless, the system will, even with external manipulation, follow some path in its phase space, a path which will not reach every point of phase space. Thus, even in this case, one is introducing fictitious states which the system will never in fact possess. As a second example, in special relativity one begins by considering all possible events (occurrences having extension in neither space nor time), which are then to be assembled into a four-dimensional space-time manifold. But

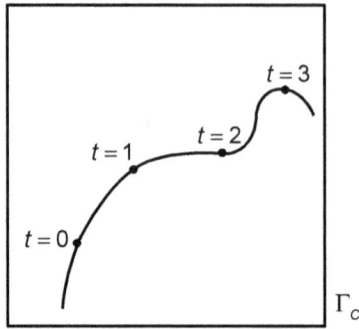

certainly there are events in space-time which no observer has witnessed. In each case, the mathematical description involves an extrapolation well beyond what is directly observed. Apparently, it is just the what we do physics to introduce, at the beginning, a framework which, if examined too closely, is unjustified.

The description of all that happens in the Universe is now in terms of the phase space of the Universe, and its dynamical trajectory. Thus, "possible occurrence" is another way of speaking of a certain region of phase space Γ_C.

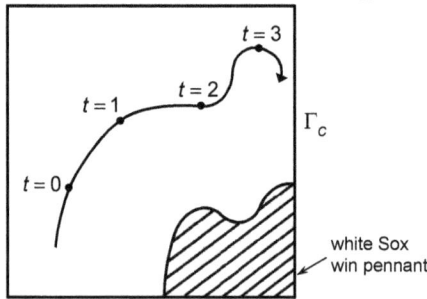

For example, "the White Sox win the pennant" defines a certain region of Γ_C, namely, those states of the Universe in which the White Sox win the pennant. The question of whether this actually occurs is the question of whether or not the dynamical trajectory of our Universe actually passes through this region. In short, we are reduced to characterizing certain regions of Γ_C and asking whether or not the dynamical trajectory of the Universe actually passes through such regions. In short, we are reduced to characterizing certain regions of Γ_C and asking whether or not the dynamical trajectory of the Universe actually passes through such regions. We emphasize that you and I (who usually think of ourselves as the external observers) are now part of the system, so our "observations" are also described, within the framework above. For example, "I make a position measurement on a free particle, and obtain $x = 2$, $y = -3$, $z = 7$" simply describes a certain region of Γ_C, through which the dynamical trajectory of the Universe may or may not pass.

One might now object that we have not, through the discussion above, acquired any additional predictive power, since we will not, presumably, succeed in finding the phase space and Hamiltonian for the Universe, much less succeed in finding our dynamical trajectory. The first point is that neither have we lost any predictive power. Our decision to regard the entire Universe as a classical system does not negate the classical mechanics which has already been done. The second point is that our goal is not to eventually discover the phase space and Hamiltonian of the Universe and predict. Rather, we are trying to push classical mechanics to an extreme to see what sort of picture emerges. The resulting picture is relatively straightforward and not particularly unpleasant. (Perhaps one might be tempted to add to this last sentence "and unjustified". One could have done the same in special relativity.)

We restate the viewpoint above in different words. Let us imagine, for a moment, an external observer O of the Universe. He has described this system by a phase space and Hamiltonian, and complacently watches it follow its dynamical trajectory in phase space. We construct a sub-system and make observations on it. O notes with satisfaction that the dynamical trajectory of the Universe indeed enters that region of phase space. We formulate some sort of objection about predictive power. O notes with satisfaction that the dynamical trajectory of the Universe enters that region of phase space. In fact, all O ever does is note with satisfaction that the dynamical trajectory enters various regions. We might as well dispense with O.

The Everett interpretation of quantum mechanics amounts essentially to the discussion above, applied to quantum mechanics rather than classical mechanics. Thus, we imagine a configuration space C for the Universe, and a differential operator H, the Hamiltonian, on C. At time $t = 0$, the Universe begins in a certain state, *i.e.*, one has a certain density ψ on C. The future evolution of ψ is then determined by Schrodinger's equation. Included within this system are all observers, their instruments, *etc.* This is the picture which emerges if one tries to apply quantum mechanics to the system consisting of the Universe as a whole. The idea of the interpretation is simply to take this picture seriously.

When referring to a system, one speaks of something happening in that system, one normally means that some classical possibility has been realized. Suppose we watch a quantum system consisting of a geiger counter and atom, such that the photon resulting from the decay of the atom would be registered by the geiger counter. Our description of this system would be by a wave function which evolves by Schrodinger's equation. The wave function may, as it evolves, become large or small for certain classical possibilities (*i.e.*, in certain regions of configuration space), but nothing ever "happens" in the classical sense. The scheme above requires that we regard the Universe as a quantum system. Thus, from this viewpoint, nothing "happens" in the Universe (in the sense that no classical possibilities are actually realized). In other words, since implicit in quantum mechanics is a wave function rather than the occurrence of classical possibilities, we are forced, if we are to regard the Universe as a quantum system, to deny the occurrence, in the Universe, of classical possibilities.

One might object, at this point, that this denial is in disagreement with our everyday observations. We, apparently, do see classical possibilities "occur". If we watch the meter attached to the geiger counter, we do see, at some point, the needle jump. The response to this objection, from the present point of view, is that this impression of ours is simply another manifestation of the wave function of the Universe. Suppose that we, as observers, were watching a quantum system consisting of some rabbits, who have geiger counters and atoms, and who do various experiments. The rabbits may indeed formulate some sort of quantum theory in which classical possibilities seem to actually occur. We, however, would describe this system quite differently. We would say that there is an evolving wave function on the configuration space of the system. Whatever impressions the rabbits get would simply be a manifestation of that wave function. Thus, from the present viewpoint, we regard our everyday impression that classical possibilities actually occur as being a manifestation of the wave function of the Universe.

One could now formulate a second objection, that our impression that classical possibilities actually occur is not being explained by this framework, in the sense that we are not shown in detail how this impression is represented by the wave function of the Universe. The point is that this is something we will not be able to explain without a great deal more information about how Humans are constructed. But this is perhaps not so unpleasant, for there are many everyday human impressions which are not adequately described by physics, presumably because of a lack of information about the workings of people.

For example, we all observe that we cannot go back and actively re-live our pasts. This observation presumably falls under special relativity. Yet it is difficult to construct some theorem in special relativity which reflects this observation. Our understanding of this everyday observation would presumably improve if we understood better the internal workings of people. Yet, one can get along quite well in special relativity without such an understanding. A similar point is to be made within the Everett interpretation. That people feel that classical possibilities are actually realized simply reflects their internal construction (that, after all, is what we would say about the rabbits). It would be nice to understand this matter better, but we have too little information, and no particular need, to do so at present.

In the Everett interpretation, no classical possibilities ever actually "occur". Classical possibilities are reflected only in the wave function is a function on the space of such possibilities (*i.e.*, on configuration space).

In fact, this link between the formalism and what people actually experience can be made slightly stronger. Consider a region of the configuration space of the Universe in which the wave function is small. We can regard this region as representing classical possibilities which "for all practical purposes, do not occur". Consider now the experiment pictured on page 91. Because of the Hamiltonian, in the region of configuration space corresponding to the atom's having decayed and the needle on the meter reading zero, the wave function is small. Thus, this classical possibility "for all practical purposes, does not occur". Similarly, if we think of the region of the configuration space of the Universe corresponding to the White Sox winning the pennant, and my impression that they did not, then one could expect that the Hamiltonian of the Universe is such that, in this region, the wave function is small. In short, one eliminates "a classical possibility occurs" as the link between the formalism and human experience. What replaces it is "in this region of configuration space, the wave function is small". In other words, one must learn to make statements about the Universe, not in the form "The following classical possibility ... occurs", but rather in the form "In the following region of configuration space ... the wave function is small". With a little practice, one can express himself equally well in this way.

We consider an example. Suppose we have some ordinary quantum system which contains a meter which, classically, can read either "A" or "B". By, *e.g.*, the Copenhagen interpretation, we determine that "the probability of reading A is 25%, while that of B is 75%". We wish to express this idea within the Everett language. The Copenhagen statement is to mean that, if this experiment were repeated many times, A would occur 25% of the time. This would not do as an Everett statement, however, because it refers to classical possibilities actually occurring.

We can, however, formulate this as follows: "In the region of configuration space corresponding to 1,000,000 of these instruments side by side, with the number reading 'A' either less than 240,000 or greater than 260,000, the wave function is small." Indeed, one would expect this to be the case. Suppose we construct a classical system which flips 1,000,000 pairs of coins, and records the number of times that both coins read heads. We quantize it, and ask what the wave function looks like after evolution through this process. One would, of course, find that the wave function will be small in the region of configuration space corresponding to "both heads" occurring less than 240,000 or more than 260,000 times. In a similar way, one formulates other statements about the Universe in terms of regions of configuration space in which the wave function is asserted to be small. Note that, whereas probability enters the Copenhagen interpretation "externally", it here enters though the internal structure of quantum mechanics itself.

In short, the Everett interpretation asks that one take quantum mechanics, as is, very seriously, and learns to live with the resulting picture. One gives up the notion of certain classical possibilities being realized in favor of the introduction of certain regions of configuration space in which the wave function is small. One carries out the same calculations, and transmits the same information, but in slightly different language. One obtains precisely the same description of the Universe that would be obtained by some external observer O. This O, however, would do nothing except look on with satisfaction as the wave function of Universe evolves. We might as well dispense with him. One does not need a classical framework in which to anchor quantum mechanics: one can just let quantum mechanics drift on its own.

Finally, one might object: "All this seems awfully philosophical and rather pointless." Imagine yourself in the following situation. You wake up one morning to discover that people always talk to each other by saying "In the region of configuration space corresponding to ... the wave function is small." That's just the way they always talk. You put up with this very confusing situation for a few days, and finally can't stand it anymore. You ask a friend to come in to see you. You say to him: "I want to reformulate quantum mechanics in such a way that classical possibilities actually occur in the Universe. I want to introduce smaller quantum systems, and observables, and breaks in the chain of instruments, on one side of which classical possibilities are actually realized. I want to modify, along these lines, the interface between quantum mechanics and what human beings actually observe. It is true that, in this program, I cannot provide details of the internal workings of people, but this feature is also common in other areas of physics." After a pause, your friend replies: "All this seems awfully philosophical and rather pointless."

About the author

Robert Geroch is a theoretical physicist and professor at the University of Chicago. He obtained his Ph.D. degree from Princeton University in 1967 under the supervision of John Archibald Wheeler. His main research interests lie in mathematical physics and general relativity.

Geroch's approach to teaching theoretical physics masterfully intertwines the explanations of physical phenomena and the mathematical structures used for their description in such a way that both reinforce each other to facilitate the understanding of even the most abstract and subtle issues. He has been also investing great effort in teaching physics and mathematical physics to non-science students.

Robert Geroch with his dog Rusty